NEW
£15.00

CW01261459

# German Artillery
## of
## World War One

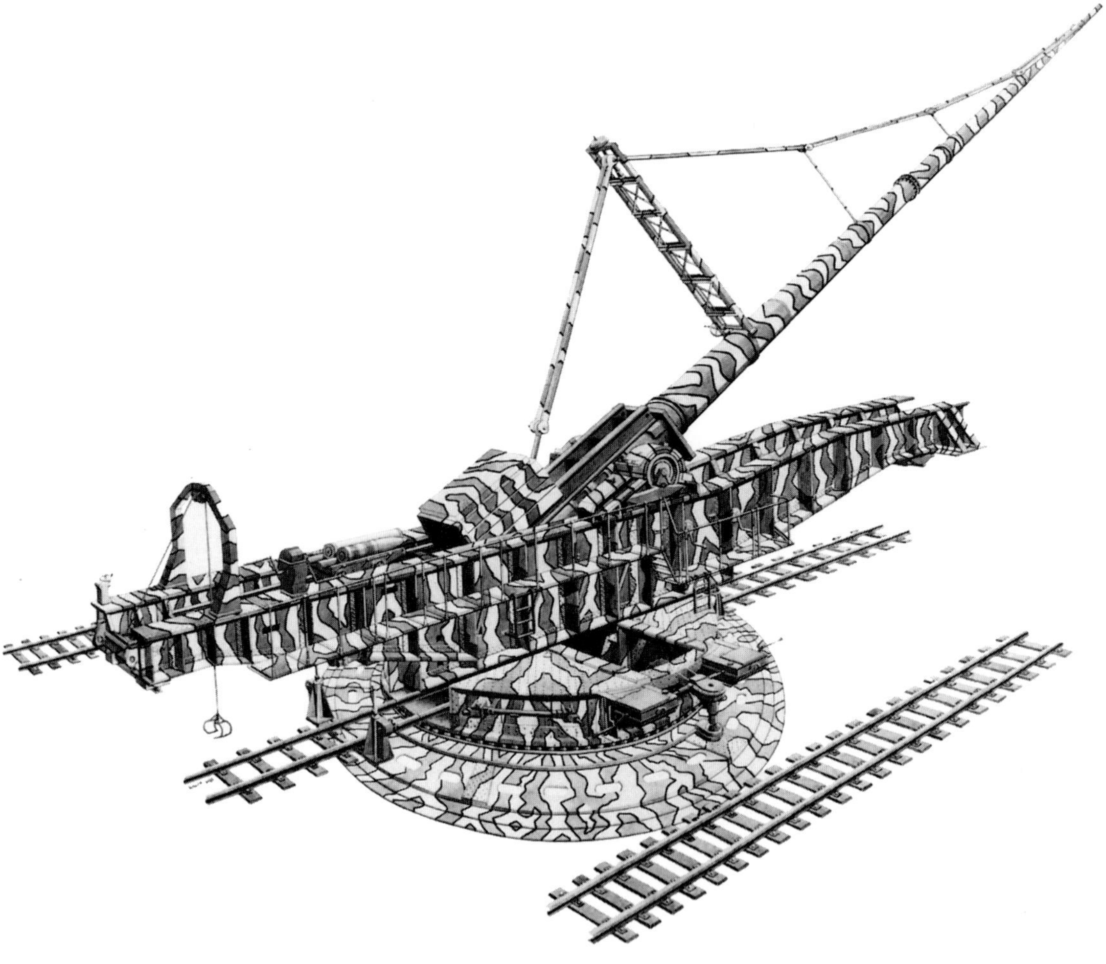

# German Artillery
## of World War One

Herbert Jäger

The Crowood Press

First published in 2001 by
The Crowood Press Ltd
Ramsbury, Marlborough
Wiltshire SN8 2HR

© Herbert Jäger 2001

All rights reserved. No part of this publication may be reproduced or transmitted in any form or by any means, electronic or mechanical, including photocopy, recording, or any information storage and retrieval system, without permission in writing from the publishers.

**British Library Cataloguing-in-Publication Data**
A catalogue record for this book is available from the British Library.

ISBN 1 86126 403 8

Photograph previous page: 21cm German Paris gun that shelled Paris in 1918 from a distance of 130km (80 miles).

**Acknowledgements**
One cannot write a book about an era almost a century ago based on one's own experiences and private archive without supporting documents. The latter I owe to many different sources, of which I want to name the following donators without forgetting my debt to the unnamed others.

- The archive of the *Gesellschaft für Artilleriekunde*, the German Artillery Study Group, was a rich treasure house of knowledge and pictures.
- The *Artillerieschule der Bundeswehr*, the School of Artillery of the German Forces, has given assistance in this venture.
- The publishing house of Bernard & Graefe kindly gave me permission to enrich my book with the pictures of their five-volume World War One classic of Muther/Schirmer, *Das Gerät der Artillerie vor, in und nach dem Weltkrieg*, published between 1927 and 1933.
- The military writer Wolfgang Fleischer also kindly helped me out with photographs from his rich archive.
- The World War One expert Dr Herman Plote widened my understanding of certain political questions and helped by furnishing me with rare photographs.
- Commander USN (ret) Charlie B. Robbins helped with information about the US investigations of the German coastal guns in Flanders in 1919.
- Captain Horst Wöhlermann provided valuable information about the German artillery at Verdun, which I received during the many times he showed me the battlefields there.

My thanks go to all of you.

Typeface used: Times.

Typeset and designed by D & N Publishing
Baydon, Marlborough, Wiltshire.

Printed and bound in Great Britain by Antony Rowe, Chippenham.

# Contents

| | |
|---|---|
| Preface | 6 |
| 1 Development of German Artillery Until World War One | 7 |
| 2 German Artillery at the Beginning of World War One | 19 |
|     Light and Heavy Field Artillery | 19 |
|     The Super-heavy Mortars, Big Bertha, Her Sisters and Aunts | 34 |
|     Super-Heavy Low-Angle Guns of the *Schwerste Flachfeuer* | 59 |
|     High-Angle Weapons of the Engineers: the *Minenwerfer* | 68 |
|     Mountain Guns | 82 |
|     Antiaircraft Artillery | 87 |
|     Fortress Artillery | 96 |
|     Coastal Artillery | 104 |
| 3 Developments of 1916 and Beyond | 115 |
|     Lessons of the War in 1914 and 1915 | 115 |
|     New Guns of 1916 | 116 |
|     German Railway Artillery | 125 |
|     Infantry Guns | 133 |
|     Antitank Guns | 141 |
|     Tank Guns | 154 |
|     Aircraft Guns | 161 |
|     The 21cm Paris Gun | 164 |
| 4 German Artillery in Battle | 177 |
|     The Battle of the Dardanelles, 1915 | 177 |
|     General von Lettow-Vorbeck and the War in Africa | 178 |
|     The Battles of Verdun | 180 |
| 5 Ammunition | 187 |
|     The Development of Ammunition until World War One | 187 |
|     Loading the Ammunition | 188 |
|     Projectiles of German Artillery | 189 |
|     German Ammunition Stocks in 1914 | 194 |
|     German Ammunition During the War | 194 |
|     New Developments of Shells and Fuses | 196 |
| 6 Tactics, Ranging, Transport and Shelters | 200 |
|     Tactics | 200 |
|     Ranging and Observation | 204 |
|     Transport | 207 |
|     Hardening the Gun Positions | 214 |
| 7 The End in 1918 | 217 |
| Glossary of German Terms | 220 |
| Index | 222 |

# Preface

On 28 June 1914, a Serbian student shot and killed an Austrian man and his wife at Sarajevo. The man happened to be Franz Ferdinand, heir to the Austrian throne. Austria, knowing the assassination to have been ordered by the head of the Serbian secret service, declared war on Serbia. Russia countered by declaring war on Austria, as did France; in turn, Germany, allied with Austria, declared war on these two countries. Britain soon joined France and Russia and suddenly almost the whole of Europe was at war. The only sensible country was Italy, which waited for the highest offer before deciding which side to join. In the end more than six million people were killed, and Austria, the lid on the Balkan powder keg, was divided into a multitude of small countries, still out for each others' blood even today. The victors, as usual the authors of history, all agreed on the culprit: Kaiser Wilhelm II of Germany, whose cry *'Gott weiss, ich habe es nicht gewollt'* ('God knows, I did not want this') has been consistently ignored.

But I am not going to rewrite history, not even that of World War One, which people afterwards referred to as 'the Great War', because they did not know that, twenty years later, there would be an even greater one. We shall look at only one side of what Lenin called 'the iron hammer of battles': the artillery of the German *Kaiserreich*. There is a lot to see, and as our understanding of any point in time means looking to history, we shall have to go back into the past sometimes.

There are many definitions of 'artillery'. One of the best, and certainly the shortest, is: 'All firearms that need to be served by more than one man belong to artillery'. This book is based on this definition.

# 1  Development of German Artillery Until World War One

Worldwide artillery had not changed for some 300 years until, in the 1840s, an Italian, with the help of a Scandinavian, served to re-ignite interest in the field. Napoleon, himself a great artillerist, had come and gone. Prussia had just introduced a new artillery system called the C/42 (C for construction and 42 for 1842, the year of introduction). This system consisted of a range of smoothbore muzzle-loading guns; the generals, and for once also the gunners, being quite content to keep their familiar type of weapons.

The only man not satisfied was a civilian. Baron von Wahrendorff's profession was to make use of the high quality of his Swedish fatherland's iron ore by casting it into gun tubes. This profitable business had been running successfully for a long time, already supplying guns to Prussia's Frederick the Great and any other customer – even opposing sides in wartime if necessary. Now von Wahrendorff turned his attention to the job of gunners, a hard one at the best of times; even the gunner of today, riding comfortably in his 155mm *Panzerhaubitze* 2000, worries that the almighty computer might fail. In those days of muzzle-loading it was even harder. On firing, the gun belched a large cloud of white smoke, jumped up and recoiled backwards for some metres. Loading the gun again was dangerous – even today those re-enacting the artillery role can be recognized by the rammer handles sticking out of their palms – but the really tough work came afterwards: the gun had to be laid again. This meant pushing the three tons of a 24-pounder (15cm/6in calibre) into the firing position and then fiddling around with levers and wedges until the sighting gun captain was satisfied and the gun laid correctly to the target. One minute later the next round was fired and then it was: 'same procedure as last time'. Having kept this up for hours, the gunners were too tired to swing their rammers in defence when the cavalry attack finally came. A gun that stayed in position would save the strength of the men who served it.

Then there was the problem of casemated fortress guns. Finding adequate space behind them for recoil travel was expensive and scarce. If the gun did not run back at each firing, a lot of room would be saved. Of course, the drawback was that the gun would no longer helpfully move its muzzle back far enough so it could be conveniently fed from the front. But this could then be done from the other end – the breech.

Breech-loading was certainly not a new idea. In fact, the first naval guns had to be set up in that way, since the invention of a suitable carriage was still pending. Of these breech-loaders, enough have surfaced to show their principle; what they hide is their problem: obturation. There was simply no way of getting a really tight seal between the massive iron beer mug forming the cartridge and the rear end of the tube. So Wahrendorff set out to invent one. In 1840 he designed a smoothbore breech-loader which closed via a piston, coating the iron balls it fired with lead in order to achieve a tight seal in the tube. From 1841 onwards, when it learned about these experiments, the Prussian Ministry of War kept an eye on them. In 1843 they ordered three of Wahrendorff's smoothbore breech-loaders, as 6-, 12- and 24-pounders.

They also observed the parallel actions of another artillery expert: the Sardinian Captain Cavalli. His

## Development of German Artillery Until World War One

idea of sealing the breech using a wedge had already been proven in the artillery of Piemont. The historic moment came when Cavalli was sent to the gun foundry of Wahrendorff in Akers, Sweden, to inspect muzzle-loading guns cast for Sardinia. He of course took his ideas along and found in Wahrendorff a kindred spirit. Together they went further, in 1846 turning from smoothbore breech-loaders to rifled ones. The principle of firing a ball-shaped bullet from a rifled barrel in handguns went back for centuries, and even the use of an elongated cylindrical-conical bullet was no longer unheard of. In 1841 Prussia had adopted the so-called *Zündnadelgewehr*, the needle gun, and in order to keep this secret had hidden it in the armouries. Unluckily, in 1848 Prussia, together with Austria and the other small German countries, finally caught on to the idea of revolutions. The people stormed the *Zeughaus*, the arsenal in Berlin, and came out laden with *Zündnadelgewehre*. On the pavements, the attachés of the European nations jostled with each other in an attempt to buy the weapons from these true democrats. So Wahrendorff and Cavalli went on with their work and in 1850 Prussia decided to join them.

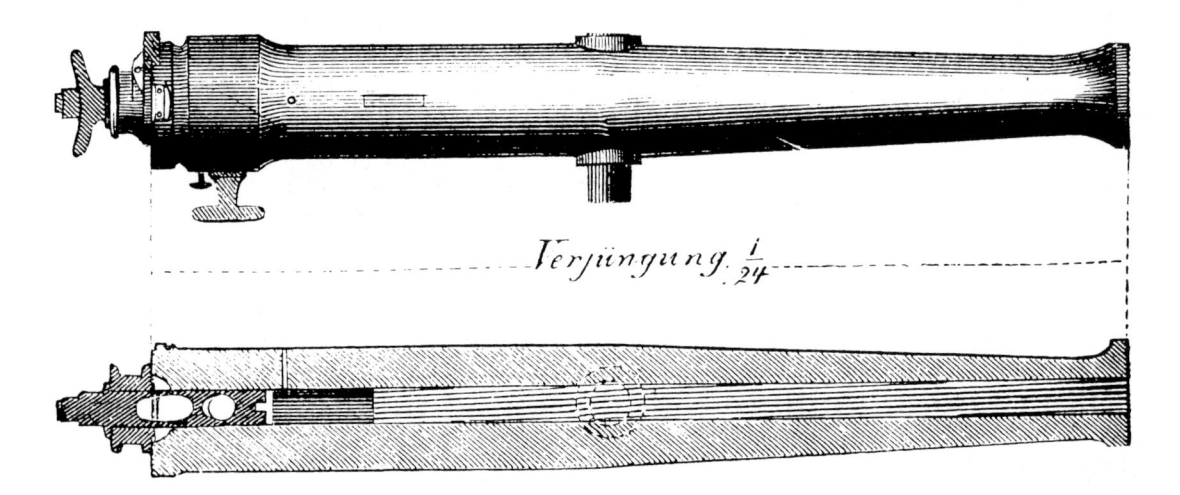

(Above) *The Wahrendorff-piston-breech system was introduced in Prussia in 1859 as the first breech-loading gun in the world. It stayed in service and was not replaced by a muzzle-loading system. In spite of the date of 1859, it was named the C/61, the construction of 1861.*

*The piston-breech system, which belongs to the longitudinal systems, as does the screw-type breech.*

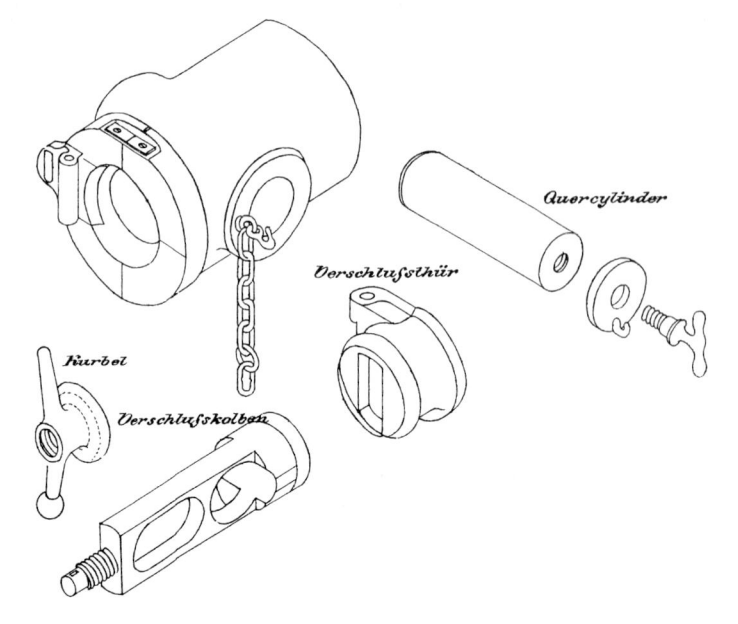

*Development of German Artillery Until World War One*

*A Prussian six-pounder (9cm) field gun C/61. The piston and the retaining bolt are missing.*

(Below) *A Belgian 12-pounder (12cm) fortress gun M 62, another Wahrendorff-type gun, captured in World War One. The breech piston and its securing pin are missing. Note the lifting rings and second trunnion rest for travelling.*

In 1851 the first Prussian rifled breech-loader – a 12-pounder of 12cm (4.8in) calibre – was test-fired. Other calibres followed and in 1854 the *Artillerieprüfungskommission* (APK), the artillery testing committee, proposed the introduction of the 12-pounder, but their efforts were in vain. Tests were resumed, mostly by breeching the walls of older fortifications, until, in 1859, two questions were answered: what material should be used for the tubes – bronze, cast iron and the very expensive new cast steel being the contenders – and who was to be armed first with these new guns – the siege trains with their heavy calibres or the field artillery with its light guns? The answer to the first question was Krupp's cast steel, which had been tested by the APK in 1855; the answer to the second question was the field artillery, which was to receive 6-, 12-, and 24-pounders, amounting to 9cm, 12cm and 15cm calibres (3.6in, 4.8in and 6in). And unlike countries such as France, with her rifled muzzle-loaders of Lahitte design taking up the spin of the rifling to the shell with the aid of *aillettes*, raised zinc warts, which were not a gastight fit and therefore lost about 10–20 per cent of the energy of the propellant, then still blackpowder, the Prussian solution was to put a lead coating around the projectile, which gave a

*Development of German Artillery Until World War One*

*Prussian 24-pounder (15cm) C/61 siege guns firing against the walls of the citadel of Juelich in Germany in 1860, to test the breaching capacity of the new pointed spin-stabilized long shells. In this contemporary illustration the attention of the spectators is concentrated on the mine being exploded in the background.*

*(Below) The next Prussian breech-loading gun type: the C/64 closed by a double wedge. Shown here is the tube of a four-pounder.*

*(Bottom) The breech system of the C/64 consisted of two wedges tightened by a screw.*

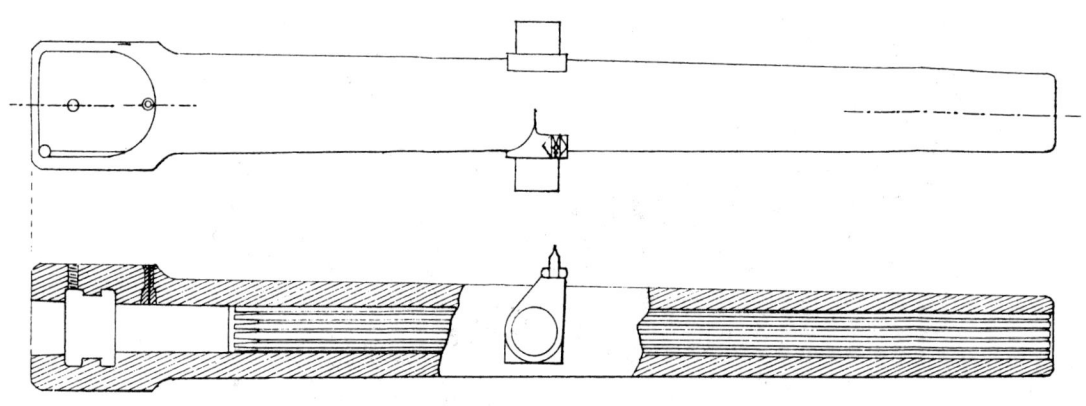

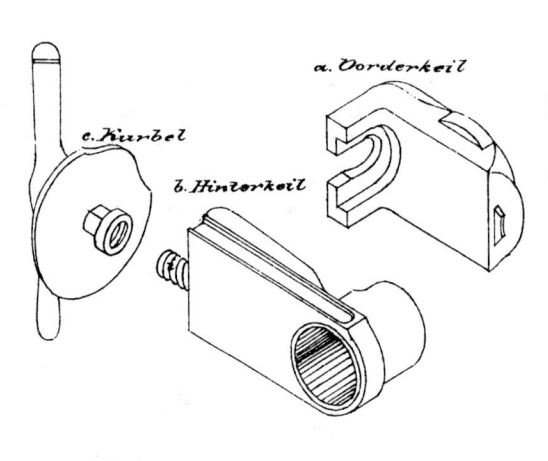

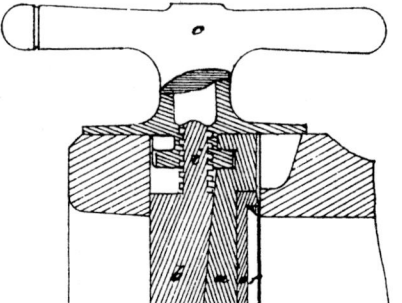

10

gastight seal inside the tube, but at the cost of 30 per cent dead weight.

The order was signed by the Prinzregent, Kaiser Wilhelm I, grandfather of Wilhelm II. He had witnessed trials of the guns and personally increased the original order of 100 blanks for the steel tubes to 300. Only these came from Krupp; the guns themselves were designed and manufactured in the *Königliche Geschützfabrik*, the Royal Gun Factory, at Spandau, a suburb of Berlin. It will remain a mystery why Prussia decided to date this first model of rifled breech-loader not after the official year of introduction, 1859, but two years later after a change in rifling, C/61.

So the Prussian artillery should have been glad that they had received the most modern guns in the world. But of course they were not. Certain problems with the obturation according to a Prussian improvement on the Wahrendorff design – an obturator pad made from pressed wood shavings – leaked gas and the gun crews complained bitterly about sooty hands. So loud were their protests that only five years later the three field gun calibres were manufactured with a completely new breechblock. This was no longer the Wahrendorff piston, but consisted of a double wedge; drawn together by a screw, the pair elongated longitudinally, thus fixing their position in the square breech end of the tube. But alas, this design of Werkmeister Kreiner, one of the foremen of the gun factory, proved to be even worse. This became apparent almost immediately, as in 1864 the moment of truth arrived: the war with Denmark over the possession of Schleswig-Holstein.

This was decided by Prussia, with Austrian aid and a great deal of help from the new rifled breech-loading guns. The 12- and 24-pounders blew apart the earthworks of the Danish fortifications at Düppeln – shades of things to come. But the guns themselves also blew apart, or at least the new C/64 model did. The wedges bent and the tubes split, with the cracking starting in the rectangular corners of the square breech opening, today recognized by any engineering student to be a design flaw. This led to a speedy return to the good old C/61, even to the extent of drilling the square breech end forged onto the rear of the new tubes for the piston breech and its cylindrical retainer pin. These were meant as temporary solutions only, but like most improvisations they became permanent, because now one war followed the other.

Only two years later C/61 fired at its Austrian brother, M61. Prussia had been generous and permitted all the German nations united in the *Deutschen Bund*, the German Union, to copy the design of the C/61. They all did, including Austria, who was a member until 1866, when Bismarck used the opportunity of Austria fighting against an uprising Italy aided by the French (not quite selflessly, as their payment was to be Nice and Savoy), to solve a problem raised by the growth of Prussia: who was to be the first, the leading nation in the German *Bund*? This was decided in Bohemia near an Austrian fortress town named Königgrätz, the battle being better known in other countries by the name of the nearby village Sadowa. The Austrians lost, thanks not so much to the modern Prussian guns – after all, the Austrians were also using them – but because of the terrible havoc wrought by the Prussian needle gun. Unlike the Austrian muzzle-loading rifles, the needle gun could be loaded in the prone position; the bearers of the Austrian rifles had to expose themselves, as the many hits to their heads showed.

The Prussian guns hardly had time to cool before they were dragged into the next war. France had cried for '*revenge for Sadowa*', though somewhat surprisingly since not a single French soldier had been involved. The French had also not relished the prospect of a Prussian prince being asked by the Spaniards to become their king. Emperor Napoleon III decided to solve this problem with his army (for which he had designed a 12-pounder gun himself, which of course was introduced, not only in France, but also in the USA, under the name *Napoleon*, maybe because the normal US citizen, not caring much for Europe and knowing even less about it, believed Napoleon I to be still running it). This war showed the Prussians and their German allies to be undergunned, this time in the rifle department. The needle gun was now almost thirty years old, and the French, unlike the Austrians, had not wasted these years. In 1866 they had introduced a needle gun of their own, of a more modern design in a smaller calibre, which outgunned the Prussians by some 300m (900ft). The Prussian artillery had to make up this deficit, their breech-loaders luckily being superior to the French rifled muzzle-loaders of LaHitte design in both range and accuracy.

But what the Prussian *Kanoniere,* gunners, really loved was a duel with the secret weapon of the French: the *mitrailleuse*. This multibarrel weapon fired salvoes of cartridges in infantry calibre, deadly against attacking infantry, but suicidal in the role in which it was mostly deployed: counterbattery fire against Prussian artillery, which outranged it by miles.

After the war, everyone concerned had learned bloody lessons. The French switched over to breech-loaders (as did all other European nations, except for Britain, which had burnt its fingers on the Armstrong breech-loader and therefore returned to the supposedly foolproof muzzle-loaders) and created a whole new line of guns with a screw-type breech, named neither after the inventor (Colonel Treuille de Beaulieu) nor the improver (Major Reffye), but after the inventor of a better obturation (Colonel de Bange). Prussia, having been united in 1871 with all of the German tribes except Austria into the second German empire, had learned two things about its artillery: firstly, at the redouts of Belfort, that it did not have heavy field artillery for knocking out stronger field fortifications; and secondly, that something evidently had to be done about the field guns.

So work that had begun before the 1870–71 war was continued, and in an astonishingly short time, given the decades it takes to finish developing a new weapon today, this gun was ready. It was introduced in 1873 – this time correctly named C/73 – and had two important changes: the breechblock was of the Krupp *Halbrundkeil*, half-round wedge, design, well proven in hundreds of large calibre naval and coastal guns already; and the smallest calibre model was made in two different sizes, which gave a light gun of 8cm (78.5mm) calibre for the dashing *Feldartillerie*, field artillery, galloping on their horses, and a heavier one of 9cm (88mm) – the birth of a famous calibre – for the dull, plodding men of the *Fußartillerie*, foot artillery, walking alongside their guns. This was accompanied by the decision to divide the artillery into these two

*Batteries of Prussian 24-pounder (15cm) C/61 siege guns firing at the forts of besieged Paris in the winter of 1870–71.*

parts – the field artillery, which had its own horses for moving the guns, and the foot artillery, which did not; in the case of manoeuvres, or even in wartime, they were supposed to hire the necessary draught animals. Fortunately, this was later changed again for the better. In 1896 the *Fußartillerie* was renamed *Schwere Artillerie des Feldheeres*, heavy artillery of the field army, and when, during a firing duel, they showed their 15cm heavy howitzers to be superior in effect to the 10.5cm light field howitzers of the field artillery, they were pressed to the bosom of the field army and called *Schwere Artillerie*, heavy artillery.

The famous *Dreikaiserjahr*, 1888, was the year in which Germany had no less than three emperors, one after another: Wilhelm I was succeeded by his

*The system of the C/73 was based on wartime experiences with both C/61 and C/64 and featured the half-round breechblock invented by Krupp. It stayed in service with the German artillery until 1918. In 1888 it was changed into model C/73/88 by lightening the tube because of less offensive smokeless powder, and again in 1891 into C/73/88/91 by the use of nickel steel for the tubes. Shown here is the six-pounder (9cm) field gun.*

(Above) *The breechblock of the C/73, invented by Krupp.*

*How the breechblock of the C/73 was drilled from the top for the ignition of the propellant by a friction wire.*

*How to teach a soldier to know his weapon? This neckerchief was for the artillery and contains all the gunner needs to know about his job, from the trumpet calls to his gun and the ammunition. Shown here is the 9cm C/73/88. There were of course also neckerchiefs for the cavalry and infantry.*

son Friedrich III, the 100-day Kaiser and husband of Queen Victoria's daughter Vicky, who after his early death from cancer was followed by his son, Wilhelm II. This is the reason why, until the 1930s, Germans used to call the British *unsere Vetter jenseits des Kanals*, 'our cousins on the other side of the Channel'. In this year the tubes of the 9cm C/73 were lightened in order to create only one field gun model, the C/73/88, and in 1891 measures taken against shells filled with modern high explosive (then *Granatfüllung 88*, picric acid) detonating inside the tube, led to the employment of nickel steel for the tubes, these guns being identified as C/91. Thus the C/73 was kept in duty into World War One.

The next generation of guns came as a surprise – and at the wrong time. Since the introduction of the C/73, blackpowder had been replaced by smokeless powder as propellant and by high explosive as filling, copper rings replaced the lead coating of the projectiles, and gun tubes were no longer made on Armstrong's multilayer principle of the *Ringkanone*, the ring gun, but of chrome-nickel steel. Time for a change, the army thought, having saved up some 20 million Goldmark for a new gun system. But they had missed an historic chance to lead again in the field of artillery. The problem with the guns was still the same: recoil. Ever since Isaac Newton had formulated his third law – *action = reaction* – the reason for the guns jumping backwards at firing, necessitating their being laid onto the target anew, was well known, but no remedy had yet been found. There had been experiments with fastening light fortress guns solidly to the walls of their casemates, like the old medieval guns fixed by dozens of boxes filled with rocks lying behind them. But this was not for field guns, destined to charge onto the battlefield, rush into firing position and hammer at the enemy troops, who all the time were keeping up fire themselves. Something revolutionary was needed … and it came.

A German engineer named Konrad Haussner, who then worked for the Krupp gun factory, came up with the idea of having the tube recoil alone, with the energy of recoil being absorbed by a suitable

## Development of German Artillery Until World War One

braking system and the rest of the gun standing still. In 1888 Haussner submitted this idea to his bosses at Krupp. They showed no interest. Haussner therefore left Krupp and, on 29 April 1881, demanded a patent on his invention.

With the help of the Prussian *Kriegsministerium*, War Department, he came into contact with Hermann Gruson at Magdeburg. Gruson was a competitor of Krupp, specializing in armour for fortifications and small-calibre quick-fire guns (until Krupp solved this problem by buying up the majority of the Gruson shares and taking over the business in 1893).

On 2 March 1894 the historic moment arrived. Haussner was able to demonstrate two guns of his design to the artillery experts of the APK. This first trial was slightly disappointing due to small imperfections in the design (the tail was too short to keep the gun stable at firing). Yet this was enough for the president of the APK to kill the future of the Haussner design for the time being. 'Take that creature away!' he demanded. And the German artillery went on to introduce a new gun, the *Feldkanone* C/96 (the C/ was dropped in 1899). This gun did not reach further than its predecessor, the C/73, since at its design the emphasis had been on trading range for mobility, so that it could follow the infantry on the battlefield. The only new feature was the metal cartridge case for the propellant, now smokeless powder.

Hardly had the army started to look with pride on its newest toy, when the French produced a shock. They had taken notice of Haussner's patents, also applied for in France, and contrary to the Krupp managers and the higher officers of the German APK, at once recognized its implications. In the year in which the C/96 went to the German artillery troops, the French produced a new 75mm gun of their own. This *soissante quinze* (75 = 60+15), as it has since been called, was truly a new gun in the sense that it was modern and had everything the C/96 was lacking, from the armour shield over the fixed cartridge ammunition to the long recoil of the tube. With fewer gunners than before, it fired more rounds per minute (approximately 25), making the C/96 look very old indeed. This was a matter not only of national prestige now, but also one of life or death, since the French kept shouting for revenge, the adventure of 1870–71 having cost them not only a few thousand million francs in a payment to Germany, money which went into the building of

*At Magdeburg, Gruson had been the world leader in the sale of armour turrets cast in Hartguss iron for coastal guns, steel cupolas based on the Schumann patents for land fortifications, and small calibre rapid-fire guns – starting with the Hotchkiss revolving gun built under licence for navy and fortresses – both for casemates of fortifications and later the movable* Fahrpanzer. *Here Gruson shows his guns in a presentation at his shooting range at Tangerhütte in the 1880s.*

*Development of German Artillery Until World War One*

*A gun built after Haussner's system of long recoiling tube, before (left), and after firing (right).*

the new German fortifications in west and east, but also the Alsace, which France had occupied in 1683 and which, with its capital Strasbourg, had returned to Germany after 200 years. So something had to be done quickly about the C/96; the solution had to be cheap, since there was no more money available.

In the meantime, Haussner had left Krupp-Gruson, as it was now named, and joined a new

*The next generation of field guns, the 7.7cm FK 96 made by Krupp, used both smokeless powder in a metal cartridge case, and, in the separately loaded shell with copper driving band, a high explosive filling of* Granatfuellung 88, *picric acid. But it still lacked a long recoil system and armour shield. Most were converted accordingly in 1904; those left unchanged were renamed FK 96 a/A (old type).*

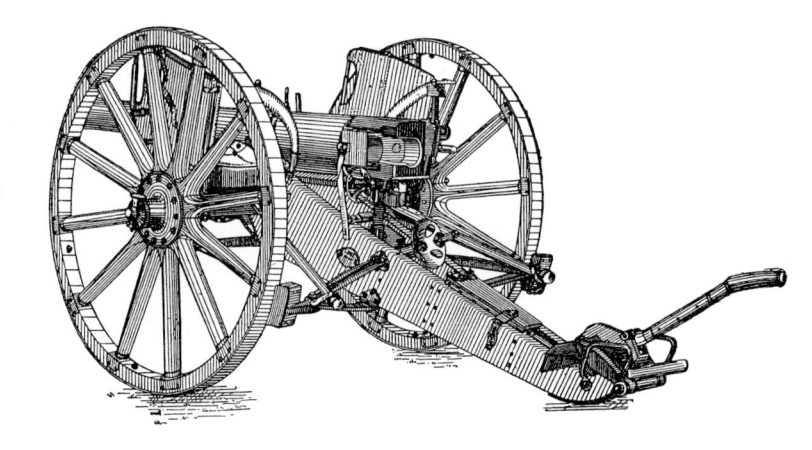

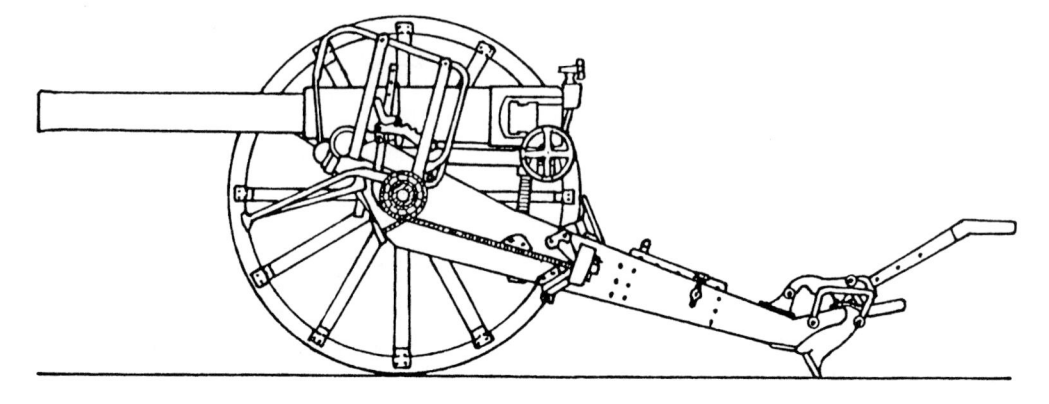

(Below) *Side view of the FK 96 showing the rope brake.*

*Krupp no longer held the monopoly on guns in Germany; for a time, Rheinmetall led the field in modern field guns with recoil system, selling them worldwide. The picture shows part of their factory at Düsseldorf, around the turn of the century.*

competitor, Rheinmetall. This had been founded by a clever man named Erhardt, who discovered that you could manufacture a steel tube (to start an artillery shell) by putting a square steel block into a round form and penetrating it with a round male die (this is still proudly shown in Rheinmetall's trademark: a square inside a circle). When Haussner came to the company in 1895, Rheinmetall had not yet made a gun, but in 1897 on the Kummersdorf range it successfully demonstrated a new gun with the long recoil, as per the Haussner principle, and by 1900 had already sold 108 of them to Britain, then at war with the Boers. Norway (132) and the USA (50) were the next customers of Rheinmetall, who also fought lawsuits with Krupp over the Haussner patents.

Finally, the APK ordered both Krupp and Rheinmetall to convert the rigid field gun C/96 into one with long recoil. This was done by reusing the old tube to save money, but turning it to a smaller diameter to save weight. Then a new breech designed by Rheinmetall was added, together with the recoil braking system of Haussner. The gun now fired fixed cartridges.

The whole updating process cost the then tremendous sum of 300 million Goldmark. To keep the whole thing more or less secret, the name was hardly changed and the gun renamed *Feldkanone* 96 *neuer Art*, new model of field gun 96; when this was written as FK 96 n/a, very few people were any the wiser. This was the field gun with which Germany would go into World War One.

But there was another German field gun, this time a field howitzer. One of the lessons of the Russian–Turkish war, especially the fighting at Plevna, was that field guns (cannon) could not cope with troops in fortified field positions. France was still discussing the problem when World War One started, whereas Germany had introduced a light 10.5cm (4in) field howitzer in 1898. This gun, the FH 98, weighed little more than the (old) FK 96 and was similarly not in possession of the modern long recoil system. Also like the FK, the FH was modernized by converting it in the same way and then calling it the *leichte Feldhaubitze* 98/09. It fired a 105mm shell weighing 15.7kg, compared to the 6.85kg of the field gun shell – more than twice as heavy.

These light field guns were supplemented by heavy field guns, heavy field howitzers and mortars. In the war of 1870–71 the Prussian artillery had fought enemy batteries by dismounting them; that is, aiming and firing point blank at the part of the enemy guns looking over their parapet. Trials in 1885 had shown that by firing in the high-angle mode this mission could be accomplished with half the number of rounds fired. This rate sank even further when modern high explosives arrived as shell filling. In 1891 it was discovered that it took a shell of about 150mm (6in) with a weight of 40kg (90lb) and a filling of 10–12kg (22–26lb) of high explosive to penetrate the cover afforded by the usual 1m (3ft) of soil on top of a layer of 0.25m (10in) wooden beams.

In 1893 Germany introduced a 150mm howitzer, also in a rigid mounting. Then the question arose as to whether this should also be converted to long recoil. Since heavy foot artillery was represented in the APK by a department (II) other than field artillery (I), which was more open to new developments, the 150mm *schwere Feldhaubitze* 02, heavy field howitzer 02, was introduced in 1902 as the very first regular artillery piece in the German army with long recoil.

Another weapon of the *Schwere Artillerie* was of smaller calibre, though similar in weight: the 10cm *Kanone* 04, the heavy 10cm (105mm/4in, to be exact) gun of 1904. This had been demanded by the *Generalstab*, the general staff, with an eye to reaching further into the rear zone of the enemy and combatting railway stations and the troops unloaded there, fortresses, and other strategic targets. From the beginning this gun was fitted with long recoil and could be transported in one load, without the tube having to be separated from the rest of the gun. Heavier cannon over 130mm calibre had to be separated. The horses then used as the only transportation by the German army were able to draw up to 4,700kg (about 10,300lb). And of course cannons, with their longer tubes, weighed more than mortars of the same calibre. So the one-piece limit was 13cm for a cannon with L/35, but 21cm for a mortar with L/12; 'L/–', indicating the performance of a gun – the longer the better with smokeless powder – is the length of the tube expressed in calibre of the gun.

There were even heavier guns in the field. In 1867, test firing had shown that the 24-pounder, at this time the heaviest calibre of German artillery, a gun of the siege trains and the fortifications, was not able to destroy certain types of fortification buildings. This led to trials with a 21cm (8in) mortar, designed by a Major Wegener of the APK. The satisfying results caused the APK to have a dozen of these built with bronze tubes and sent to Strasbourg, which was under siege by the Prussians. The impression made by the shells at the time was terrific on both the Prussian and French sides. This was followed by different modernized versions, until in 1910 the 21cm *Mörser*, the 21cm mortar, was introduced. It was a modern gun with long recoil, developed over ten years by Krupp and Rheinmetall; as with the *leichte Feldhaubitze*, they were both given contracts for its manufacture.

This ends the range of light and heavy field guns and mortars, but certainly not the list of guns developed from the 1870s onwards. But since they, like the 15cm Ringkanone of 1872, were made for land or even coastal fortifications (and were then similarly handed over as surplus from coastal to land fortifications in 1877) we shall come across them when we look at their modern versions used during the war. And *Die Dicke Bertha*, Fat Bertha, has not been forgotten. But did you know that she had a sister of equal calibre but even larger size?

*Krupp was still unrivalled where heavy guns were concerned.*

# 2 German Artillery at the Beginning of World War One

## LIGHT AND HEAVY FIELD ARTILLERY

In August 1914, when the trains crowded with soldiers rolled toward the front, everyone expected to be home again by Christmas. This was caused by the misguided expectation that this war would be like that of 1870–71, with the *Erbfeind*, the traditional enemy, France, toppling quickly. And fall quickly it would have to, because the whole planning of the German general staff had been built around this belief. It was the fault of the enemy himself, of course. Germany had enjoyed a period of security thanks to Bismarck's diplomacy; he had managed to maintain good relations with both Britain and Russia, which left revenge-thirsty France without allies.

A German field gun, 7.7cm FK 96 n/A (right), compared to a French field gun, 75mm M 97 (left), on a French postcard.

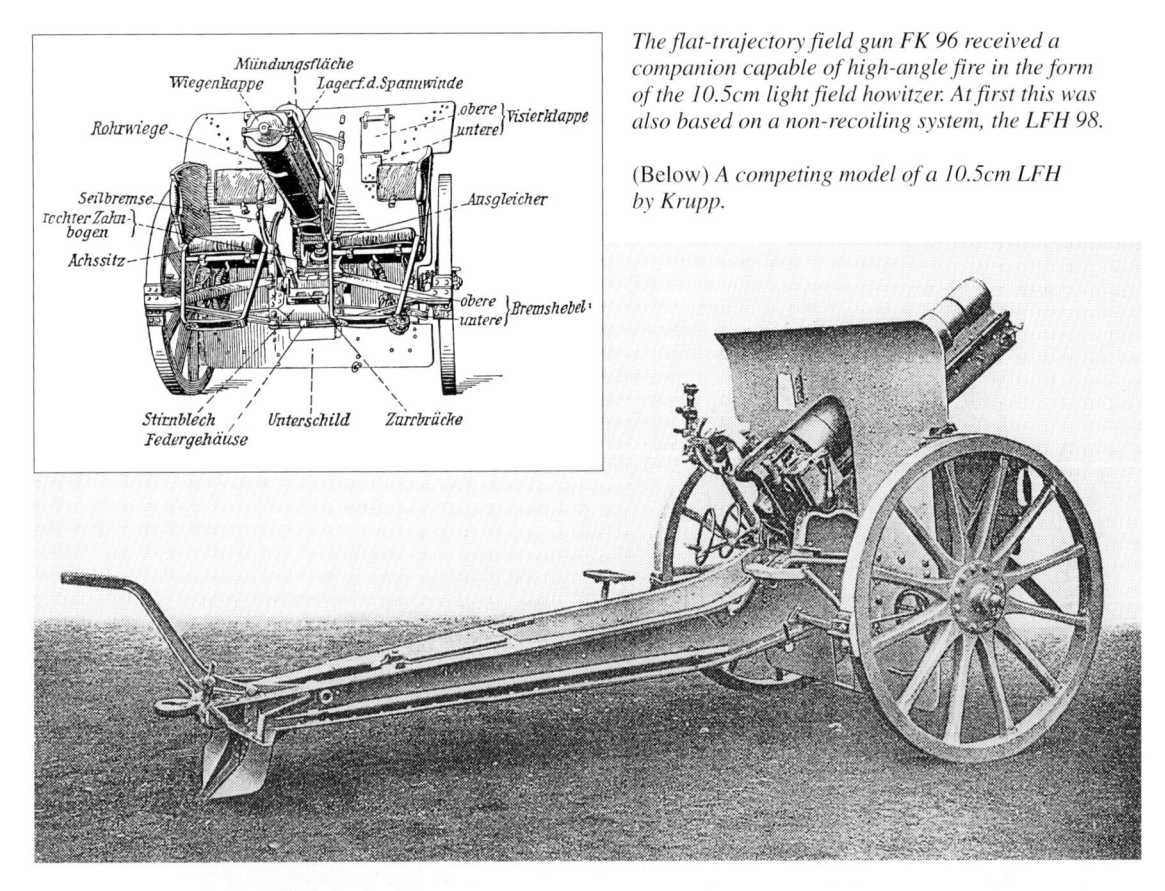

## German Artillery at the Beginning of World War One

*The flat-trajectory field gun FK 96 received a companion capable of high-angle fire in the form of the 10.5cm light field howitzer. At first this was also based on a non-recoiling system, the LFH 98.*

*(Below) A competing model of a 10.5cm LFH by Krupp.*

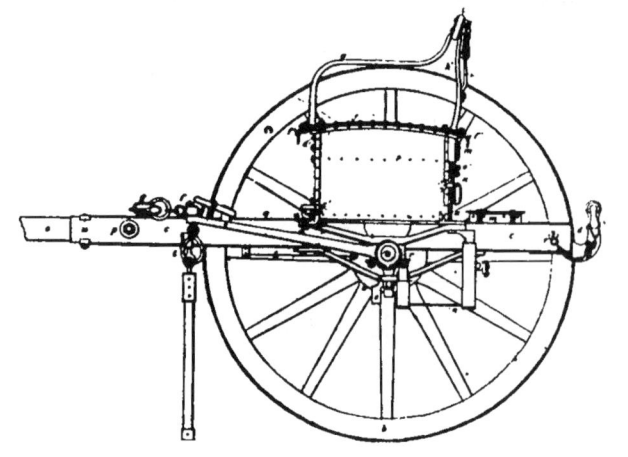

*The schwere 9cm Kanone of 1879 (left) featured a tube of hardened bronze, made after the Uchatius theories. It performed in the same way as the C/73, without a recoiling system, with which it also had the limber (above) in common.*

*German Artillery at the Beginning of World War One*

*A similar type of gun was the old 10cm gun by Krupp, introduced in 1899 and shown here in the firing position.*

*(Below) This 10cm gun by Krupp, introduced in 1902, already had a long recoil system.*

*(Bottom) The 10cm gun of 1902–3 by Rheinmetall also had a modern recoil system.*

*German Artillery at the Beginning of World War One*

*Also with modern recoil system, the later 10cm* Kanone 04 (top) *and the 10cm* Kanone 04/12 (bottom) *derived from it and fitted with a large shield to protect the top. Both were made by Krupp.*

When the young Kaiser Wilhelm II took over he was convinced that he was a better diplomat and thus Bismarck had to go. Soon France was able to form an alliance with Russia; Britain frowned at the growing German fleet; and Germany was isolated in a tight circle of enemies, with Austria its only dependable ally, in spite of the Triple Alliance signed with Italy in 1882.

Seeing Germany thus surrounded and with its back to the wall, General von Schlieffen, who had succeeded Moltke (the elder, of 1870–71 fame) as head of the general staff, hit upon the only remedy: beat the enemies one after another. First France, before the Russian 'steamroller' got into motion, and then Russia, whose mobilization was expected to take a long time. Thus the German

*German Artillery at the Beginning of World War One*

army started the long planned swing through Belgium, to reach the Channel and then press on into France, a direction of attack chosen following an evaluation of the almost impenetrable French borderline of modern, strong fortifications, all built by General Sère de Rivière after 1871. (German war theoretician Clausewitz speaking from his grave: 'If you entrench yourself behind strong fortifications you compel the enemy to seek a solution elsewhere.') This went well for a time due to the new super-heavy German mortars (*see* page 50), but the attack stopped at the Marne River, and the *Grabenkrieg*, the war of the ditches, began.

The German field artillery had not done well from the beginning. It had entered the war with 5,068 field guns 77mm FK 96 n/A, as well as 1,934

*The 10cm Krupp* Kanone *14 had a longer tube and a shield.*

(Below) *Rheinmetall's 10.5cm* Kanone *L/35 of 1913 was used as an antiaircraft gun during World War One.*

*German Artillery at the Beginning of World War One*

(Above) *The* schwere *12cm* Kanone *of 1880, shown in firing position, was without recoil system. Here it is shown on its typical 'mixed mounting' of wood and iron, necessary since this siege gun was supposed to dislodge the fortress guns of the enemy from behind a parapet high enough to protect the gun crew.*

*Neither the antiquated 15cm* Ringkanone *of 1872 (above) nor the longer and later* lange *15cm* Ringkanone *of 1892 (right) had recoil systems. They were both fortress guns, firing from behind the parapets of the annexe batteries at the enemy siege batteries.*

*German Artillery at the Beginning of World War One*

(Top) *A battery of 15cm* lange Ringkanone *92 being transported by railway.*

(Above) *The war of 1870–71 had already highlighted the need for a heavy field howitzer. This was the answer: the 15cm sFH 93, the* schwere Feldhaubitze *of 1893, a gun still without a recoil system. Captured 15cm sFH 93s in France on a French postcard.*

(Right) *A 15cm sFH 93 on the march.*

*German Artillery at the Beginning of World War One*

*German Artillery at the Beginning of World War One*

(Opposite page, top) *The next 15cm sFH came in 1902 and had a recoil system. It is shown here on the march.*

(Opposite page, middle) *A 15cm sFH 02 with L/12.*

(Opposite page, bottom) *In 1913, a new variant of the 15cm sFH appeared, named sFH 13, which due to its L/14 fired 1km further than the 02 and had an armour shield. It is shown here in firing position.*

(Below) *An sFH 13, ready to start rolling.*

*The* lange schwere *15cm FH 13 L/17, the long heavy FH 13.*

*German Artillery at the Beginning of World War One*

*The sFH 02 was updated into an equivalent of the sFH 13: the 15cm sFH 13/02, with the L/17 tube and the altered carriage of the sFH 02.*

Austrian 76.5mm FK 05, against 4,780 French 75mm 1897, 897 British 83.3mm 18 pdr and 6,278 Russian 76.2 mm FK 1902/03. On the other hand, 1,260 German light field howitzers (lFH) 105mm FH 98/09 together with 420 Austrian 104mm lFH 1899 more than matched 84 French 120mm lFH, 169 British 114mm lFH 1910 and 512 Russian 122mm lFH 1909. Later, Italy would throw her guns into the balance. The light 75mm field gun 1912 was the first gun to show a folding trail and was designed by the French Lt Col Deport, who had also co-fathered the famous *soissante quinze*.

*In 1907 Rheinmetall constructed the 15cm sFH 07.*

*German Artillery at the Beginning of World War One*

(Below) *The sFH 07 was followed in 1913 by the heavy experimental FH, the 15cm* schwere Versuchs *FH 13.*

(Inset) *The loading lever of the 15cm* schwere Versuchs *FH 13 used to seat the shell and the propellant case.*

*In 1914 Rheinmetall's 15cm sFH 14 was introduced.*

    Of the heavy field howitzers, 416 German 15cm sFH 02, together with 112 Austrian 15cm sFH 99/04 stood against 104 French 155mm sFH 1904, 86 British 152mm sFH 1897 and 164 Russian 152mm sFH 1909. Of the longer reaching and heavier guns, Germany put forward thirty-two of the 10cm *Kanone* 04, four 135mm *Kanone* and 216 of the 21cm *Mörser* 1910. The enemies were far

*Ammunition was transported in the* Munitionswagen. *The old one-part* Fussartilleriemunitionswagen *02 carrying thirty-six rounds* (left) *was superseded by the lighter two-part* schwere Feldhaubitz-Munitionswagen (below) *with the same load.*

worse off in this field. France only had old mortars of 220mm calibre, not on wheels but on a wooden platform, Britain fielded its 8-inch howitzer only after 1914; and the Russians had stowed all their guns over 152mm calibre in the fortresses, where they were either destroyed by German siege artillery or captured after the fall of the fortifications. Thus Germany showed a certain superiority in 150mm field howitzers and 210mm mortars. (When they landed in 1917, the American soldiers would arrive empty-handed as far as artillery was concerned; France would supply this.)

While German heavy firepower was superior, however, the army doctrine was not. The lessons of the last three important wars had not been learned. As we have already discussed, in the Russian–Turkish war of 1877–78, Russian artillery at Plevna had not been able to fire Turkish infantry out of their earthwork field fortifications, causing the Russians to lose 16,000 men in three attacks. The war in South Africa (1899–1902) was not taken as a sign of things to come, but as untypical of European conditions. The great tactical advantage of firing out of well-hidden gun positions was only recognized by a few, but at least the impression made by the powerful new rifles served to ensure that the new field gun 96 n/A was fitted with armour shields. Even more attention should have been paid to the Russian–Japanese war of 1904–5, the first modern war of the twentieth century. It confirmed the role of the machine gun on the battlefield and also provided a lesson in artillery. Until then, most European nations had only fought colonial wars, against enemies far inferior in terms of military training, technical equipment or numbers – or sometimes all of these. Now for the first time two enemies of equal standing clashed. The surprisingly minimal effect of the artillery was taken as atypical, the high ammunition consumption was ignored, as was the apparent inability to discover hidden batteries.

Of course there were discussions in the military magazines – after all, there had been German military observers on both sides – but everyone shied away from voicing the obvious truth: that there was no remedy against hidden guns. This would have been to say that the German war doctrine, with its emphasis on attack, was antiquated. The digging in of the infantry, the building of field fortifications, the efficiency of logistics – they all resulted in an empty battlefield in which defence was superior to attack. And all large European nations agreed that a European war would last only three months. No one could imagine such a high consumption of ammunition over such a long period.

The German field artillery would have had to draw the following lessons from these conflicts in order to go into the next war with a promise of success right from the beginning:

- give up the doctrine of offensive warfare and the cavalry mentality
- there must be effective cooperation between infantry and artillery; fire support instead of fire superiority
- improve fire from hidden gun positions
- use and improve target acquisition by sound and flash ranging
- integrate heavy high-angle guns into the field army
- enlarge the calibre of field guns
- fit the ammunition to the task of destroying cover by exchanging the shrapnel for the explosive shell as the main projectile and build up the ammunition supply per battery and gun.

Contrary to this, however, the emphasis in the training of the field artillery was on driving the guns with the horses. Small wonder then that the field artillery had to follow the cavalry regulations.

Peacetime manoeuvres mostly involved working on the change of battery from one firing position to the next, with the commander of the regiment recording the time by stopwatch. This was why the new field gun demanded and introduced in 1904 in the form of the FK 96 n/A had to be light and mobile; ballistics and terminal effect did not count. Until the war, most generals insisted that the field artillery follow the attacking infantry for as long as possible, preferably into the enemy lines, and then fire short-distance from an open emplacement. Things improved a little just before the war – but too late.

The war began on 2 August 1914, and three weeks later the first large battle was fought. The Schlieffen-based swing through Belgium and Luxemburg around Metz into the French flank had cost Germany not only politically; it had also provided the excuse for Britain's entry into the war on the side of their French allies. This raised the superiority of their combined troops at the Marne battle to double the strength of the German army, which had lost precious troops to other tasks: one corps to surround Antwerpen, one to safeguard the long supply lines and two out in East Prussia to watch the progress of the Russian steamroller. Thus they were not strong enough at the decisive point, the Marne, only some 50km (30 miles) from Paris, which seems to have decided the outcome. The German attack

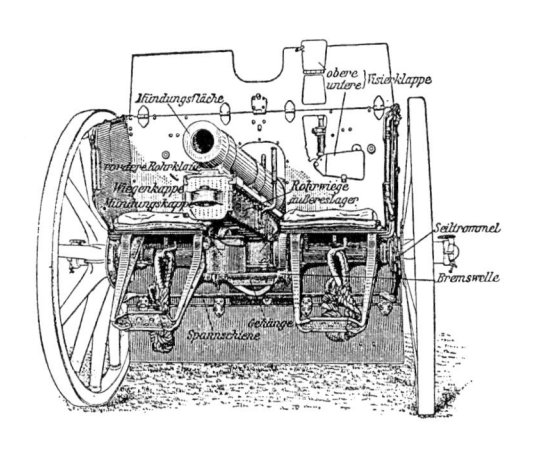

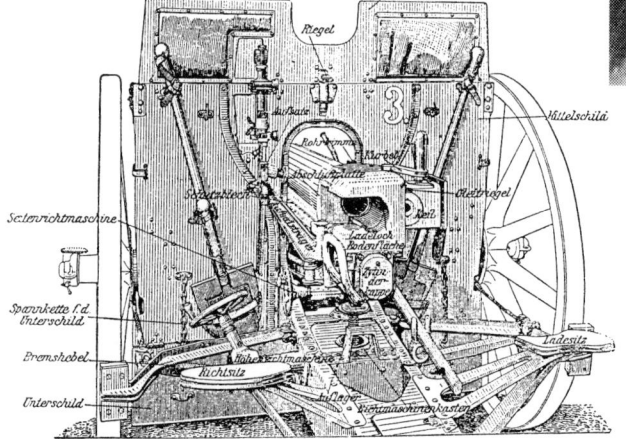

## German Artillery at the Beginning of World War One

(Above) *The 7.7cm FK 96 n/A dug into its position.*

(Left) *The 7.7cm FK 96 n/A (new type) when it was fielded in 1905. It also featured a new breechblock invented by Rheinmetall, which worked in one motion. Lack of money forced the gunners to be content with the old short tube, turned down in diameter, but now protected by an armour shield. The ammunition was now of the cartridge type and the shell filled with Füllpulver 02 (TNT). The FK 96 n/A is shown from front and rear.*

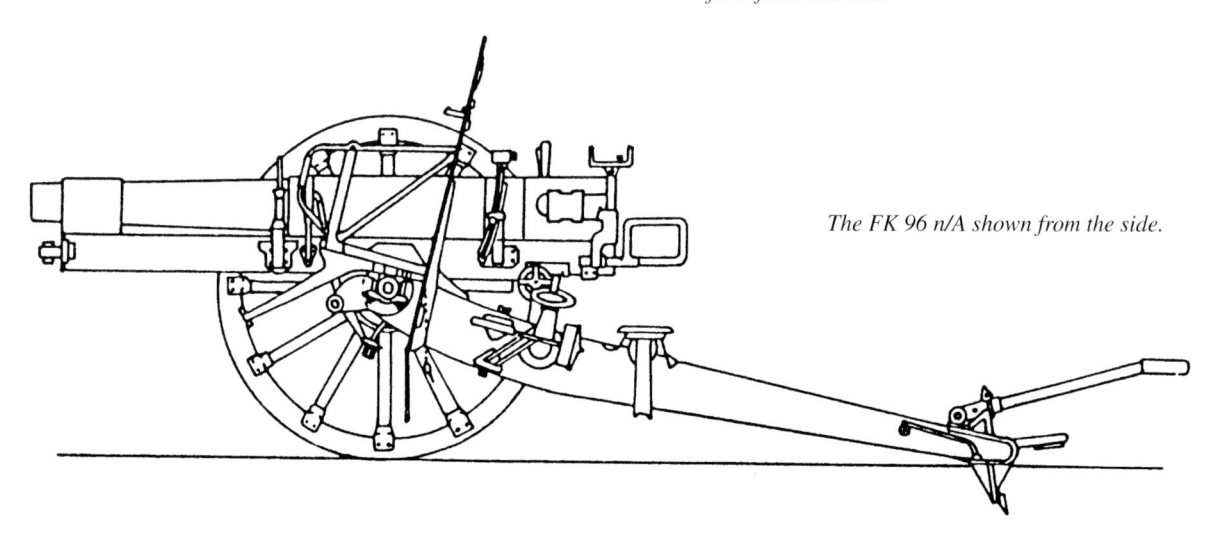

*The FK 96 n/A shown from the side.*

was stopped and never really got rolling again for the rest of the war, and now retired under the ground, until with the *Götterdämmerung* in the spring of 1918, shortly before fresh US troops and the British tank appeared on the battlefield to decide the outcome together.

The light field artillery was to stay on into the next war, with the Germans introducing another two guns of this calibre: the 7.5cm *leichte Feldkanone* 18, light field gun 18, in 1930 (more about these fraudulently dated weapons later) and the 7.5cm *leichte Feldkanone* 7M85, named using a completely new classification system and introduced in 1944. (The 7.5cm FK 7M59 is not included here, as it was simply a 7.5cm Pak 40, an antitank gun with extended elevation arc.) These 75mm guns, introduced against the ubiquitous trend towards bigger calibres, were demanded by Hitler himself, who argued – not without some merit – that a shell of 75mm was sufficient to kill a man, and how many more of them could be stored and transported compared to bigger calibres.

After World War Two, the smallest calibre seen on the battlefield grew to the 105mm field howitzer, and today, anything under the ever-present 155mm gun can only be found in the mountains, with NATO having agreed mostly on the Italian OTO MELARA model.

**Light Field Guns**

| Gun model | Calibre (in) | Weight Empl. (lb) | Tube Length (in) | Shell Weight (lb) | Muzzle Velocity (ft/sec) | Max. Range (ft) | Elevation/ Azimuth (degr.) | Remarks |
|---|---|---|---|---|---|---|---|---|
| 7.7cm FK 96 a/A | 3.1 | n/a | 84.1 | 14.4 | 1,395 | 15,000 | n/a | (1) Krupp |
| 7.7cm FK 96 n/A | 3.1 | 2,142 | 84.1 | 14.4 | 1,395 | 23,400 | +15/8 | (2) Rheinm desig. |
| 7.7cm FK 16 | 3.1 | 2,783 | 108 | 14/12.4 | 1,635–1,806 | 32,100 | +40/4 | (3) (4) Krupp |
| 10.5cm lFH 98/09 | 4.2 | 2,572 | 67.2 | 33.2 | 906 | 18,900 | +40/4 | Krupp |
| 10cm K 04 | 4.2 | 5,895 | 125 | 39.4 | 1,680 | 38,100 | +30/4 | Krupp |
| 10cm K 14 | 4.2 | 5,909 | 147 | 39.4 | 1,755 | 39,300 | +45/6 | Krupp |
| 10cm K 17 | 4.2 | 6,825 | n/a | 39.4 | n/a | 42,300 | +45/6 | (5)Krupp |
| 10cm K 17/04 | 4.2 | 6,615 | n/a | 39.4 | n/a | 42,300 | +30/4 | Krupp |
| 9cm K C/73/91 | 3.5 | 2,747 | 84 | 15.75 | 1,326 | 19,500 | n/a | (6) Krupp |
| 9cm K 79 | 3.5 | 2,747 | 84 | 15.75 | 1,326 | 19,500 | n/a | (6) (7) Krupp |
| 10cm K 99 | 4.2 | 5,544 | 126 | 39.4 | 1,653 | 38,100 | +35/? | (6) Krupp |
| 10cm K 02 | 4.2 | 5,880 | 125 | 39.4 | 1,680 | 30,900 | +30/4 | Krupp |
| 10cm K 02/03 | 4.2 | 5,880 | 125 | 39.4 | 1,680 | 30,000 | +30/4 | Rheinm |
| 10cm K 04/12 | 4.2 | 5,895 | 125 | 39.4 | 1,680 | 38,100 | +30/4 | Krupp |
| 10cm Küst K L/50 iRl | 4.2 | 14,300 | 210 | 35.2 | n/a | 58,500 | n/a | (8) Krupp |

Also used captured guns: old Russian 10.67cm K 77; Italian 10.2cm motorized guns.

*Remarks:* (1) FK 96 alter Art, separate loading with metal case, only rope brake for recoil; (2) n/A made from a/A, new breech system firing cartridges; (3) separate loaded; (4) 96-/C-shell; (5) transported in two loads; (6) old gun without recoil system; (7) like C/73, but with a tube of hardened bronze; (8) former coast gun.

n/a = no data available; Rheinm = Rheinmetall.

## Heavy field guns

| Gun model | Calibre (in) | Weight Empl. (lb) | Tube Length (in) | Shell Weight (lb) | Muzzle Velocity (ft/sec) | Max. Range (ft) | Elevation/ Azimuth (degr.) | Remarks |
|---|---|---|---|---|---|---|---|---|
| 12cm sK 79 | 4.8 | 5,155 | 112.4 | 35.7 | 1,317 | 23,700 | +40/? | (1) Krupp |
| 13cm K 09 | 5.3 | 14,360 | 189 | 88.2 | 2,085 | 49,500 | +26/4 | Krupp |
| 15cm sFH 93 | 5.9 | 4,274 | 64.8 | 85 | 840 | 18,150 | +65/? | (1) Krupp |
| 15cm sFH 02 | 5.9 | 4,274 | 71 | 85 | 975 | 22,350 | +42/4 | (1) Krupp |
| 15cm sFH 13 | 5.9 | 4,641 | 83.6 | 85 | 1,095 | 25,500 | +45/5 | Krupp |
| 15cm VersH L/30 | 5.9 | n/a | n/a | n/a | n/a | n/a | n/a | (2) Rheinm |
| 15cm Ring-K 72 | 5.9 | 10,353 | 138 | 88.2 | 1,188 | 23,700 | +37/? | (1) Krupp |
| 15cm lange K 92 | 5.9 | 12,667 | 179 | 92.4 | 1,449 | 36,000 | +40/? | (1) Krupp |
| 15cm K i.S.L. | 5.9 | 25,452 | 238 | 108 | 2,064 | 58,500 | +30/360 | (3) Krupp |
| 15cm sK 16 | 5.9 | 21,294 | 245.6 | 108 | 2,247 | 68,400 | +42/8 | (4) Krupp |
| 15cm sK 16 | 5.9 | 19,110 | 245.6 | 108 | 2,247 | 68,400 | +42/8 | (5) Rheinm |
| 15cm SK L/30 iRL | 5.9 | n/a | 180 | 108 | n/a | n/a | n/a | (6) Krupp |
| 15cm SK L/40 iRL | 5.9 | n/a | 240 | 108 | n/a | n/a | n/a | (6) Krupp |
| 17cm SK L/40 iRL | 6.8 | 49,350 | 272 | 134.4 | 2,445 | 72,000 | +45/8 | (4) (5) Krupp |
| 18.5cm Haubitze | 7.4 | 10,867 | 163 | 168 | 1,320 | 33,000 | +45/5 | (6) Krupp |
| 21cm Mörser 99 | 8.4 | 10,143 | 84.4 | 174.3 | 1,182 | 24,600 | +70/? | (1) Krupp |
| 21cm Mörser 10 | 8.4 | 15,496 | 101.2 | 252 | 1,101 | 28,200 | +70/4 | Krupp |
| 21cm Mörser 16 | 8.4 | 15,866 | 122 | 252 | 1,182 | 30,600 | +70/4 | Krupp |

Also used captured Belgian 15cm sK; Russian 15cm sK and 15.2cm sFH; British 12.7cm K and 20.3cm/8in sFH short and long; Italian 21cm mortars and 30.5cm mortars; Belgian 21cm mortars; French 22cm mortars, 30.5cm and 37cm railway guns. German 15cm guns had a true calibre of 149mm.

*Remarks:* (1) old gun without recoil system; (2) also called *leichte Kartaune*; (3) mobile ex-fortress gun in armour housing; (4) transported in two loads; (5) transported in one load, hung from *LastenverteilerG*; (6) SK = ex-naval gun; (5) on wheels, later on railway; (6) tube in two parts, only twelve howitzers made.

n/a = data not available; Rheinm = Rheinmetall.

# THE SUPER-HEAVY MORTARS, *BIG BERTHA*, HER SISTERS AND AUNTS

The road through Belgium to the coast of the Channel was not an easy one, because Belgium had built certain obstacles along the way. Soon after its constitution in 1831 out of the southern part of the Netherlands – with the help of French and British warships sailing up the Schelde and bombarding the Dutch citadel in Antwerpen – Belgium had become a prosperous country, situated between two large hungry neighbours. So the Belgians had to protect their wealth by fortifying their frontiers. This they did twice before World War One, having the good fortune to be able to claim as one of their own one of the greatest experts in modern fortress-building.

Henri Brialmont had, as a young captain in the 1850s, protected Antwerpen via a circle of large forts around it, and later repeated this in the 1890s at Lüttich (German)/Luik (Flemish)/Liège (Wallonian French), Namers/Namur, and again at Antwerpen/Anvers. These formed the basis of Belgian war doctrine, which (simplified) ran like this:

if our German neighbours attack, we shall stop them at Liège and wait for the French to arrive. If our French neighbours attack, we shall hold them up at Namur until the Germans come. And if the two should attack simultaneously, we retire into the national redoubt of Antwerpen and wait for the British to come to our aid.

In 1914, the German *Kaiserliche Armee* came. Lüttich is only half an hour's drive (about 60km/35 miles) on the Autobahn from the German border-town of Aachen. But around the town an impressive ring of strong forts cried a loud and firm 'Halt!'. They were of modern design, built in concrete and well armed with Krupp guns of up to 21cm (8in) calibre in French armour turrets. Crossing the frontier on 4 August the Germans (25,000 men; later 90,000 with 90 pieces of artillery and 108 machine guns) soon met the defenders of Liège, who numbered 31,900 men with 252 fortress guns and 30 machine guns. After the fighting started it became apparent that both sides were unused to what Clausewitz called 'the frictions of war' – you never know exactly what the enemy is planning or doing, where your neighbours are, and so on. Thus during the Belgian retreat one German General von Ludendorff drove to the old citadel, which he took to be in German hands already. When he discovered it to be still in Belgian hands, he demanded that the defenders surrender, which they did.

This unbloody warfare did not work with the strong modern forts, however, which needed other measures. The mobile 21cm mortars opened fire on some of the eastern forts of the girdle on 5

*The 21cm mortar fired for the first time at the siege of Strasbourg in the German–French war of 1870–71. The first clumsy models finally led in 1899 to a weapon looking much like its contemporary sister, the 15cm SFH 93. But the higher weight of the mortar forced it to travel in two loads.*

*German Artillery at the Beginning of World War One*

August. Three days later, the first (Barchon) surrendered, followed three days later by Evegné. At this rate it would have taken months for the whole fortified circle to capitulate, if at all. But this was to change.

On the evening of the 12th, two super-heavy 42cm (16.5in) mortars arrived. They had been loaded onto the railroad at the Krupp factory at Essen, transported by tractors of the famous Sarrasani circus and the private horses of Krupp's son-in-law, von Bohlen-Halbach. This combination moved over 42 tons of the heavy pieces (with the gun in position: 42,600kg/92,000lb) into their first firing positions. They opened fire at Fort Pontisse that evening. At

*When the guns received recoil systems, the 21cm mortar was also altered. This led to a first experimental version of 1907: the 21cm* Versuchsmörser *L/10 (above), shown here with its tube separate on the tubecart (below).*

# German Artillery at the Beginning of World War One

*The L/12 version simply named* Der Mörser, *the mortar, shown here with tube lowered* (top) *and tube on cart* (bottom).

(Below) *Here the tube of* Der Mörser *on its cart is already pulled by motor traction, as demonstrated on a French postcard.*

*German Artillery at the Beginning of World War One*

(Opposite page)
*Rheinmetall also developed their version of a 21cm mortar, firstly in 1907 with L/12* (top)*, and later in 1909 with L/15* (middle)*. Neither was adopted.*
(Bottom) *12cm* Versuchsmörser *L/12 (Rheinmetall) pulled on elongated pedrails.*

noon next day (13th) the fort drew up the white flag, as did two others. On the 14th another two surrendered, and the same on the 15th. Then the 42cm mortar fired at Fort Loncin. The fifth 810kg (1,700lb) shell hit the fort in a place in which an earlier hit must have formed a crater in the concrete. The shell penetrated the 2.5m thick roof of the fort and detonated inside, right in the middle of an ammunition magazine. This of course blew up, taking the right half of the fort with it and killing 350 men of the garrison, who still lie buried under the ruins. At this moment the German infantry stormed the ruins and took prisoner the Belgian General Lehman, who was leading the defence of Liège and had transferred his command post from the citadel to Fort Loncin.

The remaining forts saw this as a good reason to surrender, and the grateful German infantry, glad that they would not have to run attacks against them, gave their Krupp gun a name of honour, which was derived from the first name of Bertha von Krupp und Bohlen-Halbach, the only living bearer of the world-famous Krupp name. The name given to the gun – *Dicke Bertha* (Fat Bertha) was to become increasingly glorified during the course of the war. The official name of the gun was either 42cm *Mörser in Radlafette* (42cm mortar on wheel carriage) or *kurze Marinekanone* 14 (*M-Gerät*) (short naval gun 14 (M equipment); the M stood for *Minenwerfer*, mine launcher). And as before this gun there had been a 42cm *Mörser* L/16 (42cm mortar) or *kurze Marinekanone* 12 (*Gamma-Gerät*) (short naval gun 12 (Gamma equipment)) introduced in 1912, and even earlier two 30.5cm *schwerer Küstenmörser (Beta-Gerät)*, M 97 L/8 and M 09 L/16, things do start getting complicated. So it is time to tell the story of these wonder weapons from the beginning.

This beginning was over fifty years previously, in a war in which Germany was for once not involved; in fact, her present enemies fought this one out between themselves. It was the Crimean War of 1854–56. This saw the first appearance of (French) swimming armoured batteries before Sevastopol, whose iron plates protected them against Russian coastal artillery. This led to a race of armoured warships, wherein each nation kept rivetting thicker and thicker iron plates to the sides of their warships. The gunmakers answered in the form of larger and larger calibres, first in muzzle-loading smoothbores such as the US columbiads of 15in and even 20 in, and then the rifled breech-loaders, with Krupp leading his competitors with a 42cm coastal gun.

In the 1890s this leapfrogging showed signs of going on forever; it was time to start on a new line of thinking. Clever observers had noticed that the warships had gained weight around the middle. The decks of the ships had not thickened in the same way, simply because the ships would then have had trouble staying afloat, burdened as they were already. So coastal batteries and forts were no longer armed solely with cannon firing with flat trajectories, but also with mortars firing in their usual high-angle mode. When their shells were dropped onto the still thin decks of ships they would penetrate them and sink the ship – a truly wonderful idea, so everyone thought. It also worked fine, as tests showed. But the problem was not one of penetration, but of hitting the ships. With flat trajectory guns you had a greater chance of a hit: imagine firing a rifle through a dozen hoops standing upright – no problem. But try this in the high-angle mode, by throwing a grenade into a basketball basket. Then you hit the *Waffenstreuung*, the dispersion of the weapon, to which you have to add that of the ammunition, and so on.

Let it be sufficient to say that within another ten years the high-angle fire against naval targets was dead. Thus all the military in the foreign countries around Germany – now no longer good friends – either smiled politely behind their hands, or openly, when they learned in 1897 that Germany was

*German Artillery at the Beginning of World War One*

going to introduce a new *Küstenmörser*, a coastal mortar. It was to be of 30.5cm (12in) calibre with an L/8 tube, and was followed in 1909 by a longer model with an L/16, both named *Beta-Gerät* (there had been an *Alpha-Gerät,* too), and in 1912 by another coastal mortar or naval gun, as this was called, in larger calibre and therefore called *Gamma-Gerät*. It was still just another obsolete high-angle coastal mortar, however, that no one but the Germans put money into any longer. In 1914 just nine of the old M 97 and only two of the M 09 existed and no more were built except one 30.5cm howitzer L/17 on wheels in 1912.

However, neither Beta nor Gamma were intended for coastal defence. They were super-heavy mortars designed to be mobile for land transport to the new fortifications France had built along her frontiers after 1871, a result of tremendous effort and expense. This had been there to see before, but on that occasion it was in another country – Russia. There the Japanese, in the war with the Russians of 1904–05, had concentrated on destroying the Russian fleet. This was not planned in the classic way by meeting it with their own fleet and then slugging it out broadside by broadside; this had already been tried and now the remains of the Tsar's fleet were hiding out in the naval base of Port Arthur under the watchful guns of their coastal batteries, not an easy prey in those days of rudimentary fire control on swimming gun platforms, when one gun on land was considered

*Next came another large calibre mortar: the 30.5cm* Küstenmörser *L/8, called the* Beta-Gerät, Beta-equipment. *The Beta-mortar is shown left, with the bedding transported on a* Lastenverteilergerät, *load distributor, below.*

(Opposite page) *Before the war this 30.5cm Beta-mortar was fielded in two batteries built for transport by railway* (top), *and another two in 1917 designed for road transport, of which the one with the carriage is shown* (middle), *being loaded onto a railway car.*

(Bottom) *The 30.5cm* Beta-Gerät *also had two parts, later slung between the fixings of a* Lastenverteilergerät.

*German Artillery at the Beginning of World War One*

*Lengthening the tube from the original L/8 by 100 per cent to L/16 gave the next generation of Betas: the* schwere Küstenmörser 09 *or* Beta-Gerät 09. *Note that with the 30.5cm Beta, Krupp had turned away from their traditional lateral breech of sliding wedge-type and used an axial screw-type breech, like almost all other gunmakers worldwide. And despite the fact that this breech did not need it for obturation, the propellant was contained in a metal cartridge case, seen here on the crew platform, for speed of loading.*

(Opposite page) *The next step in the calibre race was the 28cm* Küstenmörser, *coastal mortar, first with L/12 (top) and later L/14 (bottom).*

to be equal to a warship at sea. So the Japanese dragged their 28cm coast mortars (or howitzers, of which it was rumoured that they had bought one from Krupp and copied this a hundred times) by sheer effort of the gunners up the steep hills surrounding the harbour and then started firing.

First, the surprised Russian warships lying at anchor at known, fixed distances were hit, but the gunners did not yet tire of their fun. The next targets were the forts. It will not astonish anyone today to learn that the modern semiarmour-piercing shells designed to penetrate the iron decks of ships also penetrated the masonry tops of the 'bombproof' casemates of the fortifications. This term was already antiquated, however, guaranteeing safety only when the old round mortar bombs of the smoothbore muzzle-loader days were dropped onto them. Then the usual 2–3m (3–6ft) of soil would soften the relatively small shock of their exploding blackpowder filling and the vaults underneath this the rest.

Now things had changed: the pointed hardened steel ogive of the spin-stabilized elongated shell with an L/2–4 fired by the rifled breech-loaders penetrated both earth and masonry, and the delayed action of their base fuses detonated the powerful modern high-explosive filling of both picric acid and TNT with a detonation velocity of about 7,000m/sec (21,000ft/sec), more than twenty times that of good old blackpowder with its slow 300m/sec (900ft/sec), and blew everything up in a high fountain.

This was the birth of the super-heavies destined for battling modern fortifications built of concrete, something the Danish had started in their Copenhagen forts built in the 1850s, even if only for reasons of waterproofing them against the Baltic Sea.

Now the concrete, with its addition of broken bricks, had been improved by the addition of sand and gravel and the insertion of a firm skeleton of reinforced steel bars, a much tougher nut to crack. This was also true of the guns of the fort, now no

*The 28cm* Küstenmörser *also had to travel in two loads: the tube on its cart* (top) *and the carriage* (bottom).

longer standing on an open rampart, but hiding underneath an armour cupola of 30cm (1ft) of nickel steel.

Thus in Germany work began to give to these mortars the thing they still lacked: mobility. At that time, no heavy coastal gun could leave its place of manufacture and be transported in one piece to its emplacement. After the inspection of the buying military it had to be broken down into smaller portions, which could be transported by railway car to its final position and rebuilt there for good. The new objective was to make these portions, 'loads', suitable for a quick dismount, give them their own special transportation and the necessary lifting means of heavy cranes, and have them rebuilt quickly, not just once but every time they were needed somewhere else. This happened to the 305mm *Küstenmörser* L/8 introduced in 1897. It was an old design, without the tube recoiling inside a cylindrical jacket in the modern way, which the genius of Mister Vavasseur, chief designer of Armstrong's, had invented, but in the old way, with the tube running up an incline of the carriage during recoil, to return to battery position by gravity. Now it really was modernized. The length of the tube now recoiling inside a jacket was doubled to L/16, the mortar and its crew were both protected by an armour shield, and the weapon was no longer laid out for transportation on light rails, but by motorization. And the most apparent innovation was its breech, no longer the traditional Krupp sliding wedge, but a screw-type of Krupp-Welin design (1891), almost unique in German artillery. We should just take a quick look at the advantages of these two different breech systems.

Ever since the introduction of the C/73 into German field artillery, the army had stuck with the *Keilverschluss*, developed by Krupp to world fame. This firm had been able to combine relative ease of manufacture with absolute strength and safety in their design and saw no reason to change a winning formula. Of course they kept well informed about modern developments and even tried them out, for example the highest refinement of the screw-type breech: the triple staged breech of the graduated Norwegian engineer Welin, whose breechblocks even today close the breech ends of the 16in guns of the *Iowa*-class battleships. But in the end Krupp always returned to their trademark *Rundkeil*, the round wedge, which in truth was only half round, the frontal side being flat. It was moved by a screw, the *Leitwelle*, the guiding spindle, horizontally through an equally shaped recess cut in the breech end of the tube. In later years the wedge grew into a rectangular prism, and on its left side was cut a semicircular opening, the *Ladeloch*, the loading hole, which was supposed to save the ramming hand of the loading gunner in case the person who closed the breech was a bit faster than him. But otherwise Krupp's breeches remained the same, right up to that of the biggest cannon (not artillery piece) ever built: the 80cm (32in) *Kanone* DORA, a railway gun of World War Two.

By choosing the wedge-type breech, the German military also unknowingly voted for a serious logistical problem. No one was then able to predict that in the course of the following decades and centuries the cost of killing an enemy by artillery fire would increase tremendously, in numbers of shots fired and material used. In the days of sooty old blackpowder fire this meant powder and shot or shells. But by around 1880, in the days of modern small-calibre quick-fire guns with their fixed cartridges, it also meant brass for the cartridge cases. And almost simultaneously with Krupp exchanging the obturation of their breeches from the steel or copper Broadwell-rings (despite his British name he lived at Karlsruhe in Germany) to the self-obturating brass cartridge case, this problem spread to the large calibres too. Thus in World War One, with its tremendous hunger for ammunition, the problem of getting enough copper for brass to make the driving bands on the shells and the cartridge cases for the propellant was enough to make technicians and logisticians weep. Thus Germany became the worldwide master in finding substitutes for all sorts of raw materials.

But in the 1880s, they were still in a blissful state of ignorance, and only one exception to this inflexible rule of the *Keilverschluss* had been discovered so far: the 15cm (6in) *Mörser* of the *Fußartillerie* in 1882, a weapon destined to help defending fortresses by lobbing big shells into the trenches of the attacking sappers, with a non-recoiling bronze tube, fortified by the autofrettage (by pulling through pistons with increasing diameter) invented by the Austrian General Uchatius, the same man who first thought about dropping bombs from an aircraft. In this case it was a balloon and the cast-iron bombs were dropped in 1848 on Venice. The reason for this choice seems to have been fathered by the desire to better the interior ballistics, and the same to have influenced the breech design of the second *Beta-Gerät* of 1909. This was transported in five loads by motorized transport or railway, but it was still a *Bettungsgeschütz*, a bedding gun, resting on a firm foundation, which had to be excavated and filled with concrete that took one week or longer to harden. Only then were you able to fire the first round. It took a very obliging enemy to

*German Artillery at the Beginning of World War One*

wait this long without bombing the new firing position with his artillery in the meantime.

But in 1909 no one thought about this, and the APK, just to be on the safe side, decided to go one bigger and therefore better and – as a sort of parallel development – scaled the 30.5cm mortar up to 42cm (16.8in), with L/16 tube, screw-type breech, concrete foundation and all. This follow-up to the *Beta-Gerät* was called *Gamma-Gerät*. The weight of the mortar had grown from 46,000kg (99,000lb) for the Beta to 150,000kg (330,000lb) and Gamma fired shells weighing up to 1,160kg (2,500lb) compared to the 410kg (900lb) feathers of Beta.

Now it took ten railroad cars to transport one of these mortars, broken down into suitable loads.

This *kurze Marinekanone*, short naval gun, as it was also named for camouflage, was introduced in 1912, and by the beginning of the war five of them had been supplied to the troops; only a select few knew about this top secret weapon, of which ten were manufactured by Krupp. One survived 1918 and the *Internationale Militärkontrollkommission* (IMMK), the International Military Control Commission on the Krupp range at Meppen, and was recommissioned for the shelling of Maginot line *ouvrage* Schoenebourg in 1940.

But this was not the end. Someone in the APK should have realized that this giant was a bit unwieldy for modern war. So Krupp should have been asked to scale this Gamma down a bit, saving

*The next step was enlarging the calibre again, which led to the 42cm* kurze Marinekanone *L/16 of 1912, short naval gun, as it was named in order to hide its true purpose: to crack fortifications. It was also called* Gamma-Gerät. *It too was a bedding gun, weighing 150 tons and transported on no fewer than ten railway cars, of which the six for the gun are shown below. The other four were for the bedding.*

| Bettung | Kran u. Zweibein | Pivotsockel | Lafette | Wiege | Rohr | Plattform m. Winde Geschoßaufzug u. Geschoßwagen |
|---|---|---|---|---|---|---|
|  | 7 000 kg | 12 500 kg | 20 000 kg | 15 000 kg | 21 900 kg |  |

*German Artillery at the Beginning of World War One*

*Installing the gun started with the unloading of the bedding parts ...*

*... followed by the mounting ...*

*... and finally the tube weighing 22 tons.*

*German Artillery at the Beginning of World War One*

*Then the* Gamma-Gerät *was loaded: its shell, cartridge case with propellant, and its screw-type breechblock swung shut and turned closed.*

*Half of the* Gamma-Mörser *were later protected by an armour housing like that of a naval gun, which they pretended to be.*

*42cm* Gamma-Mörser *in action in the West.*

on the tube length and on shell weight too, in order to place the gun on a wheel carriage, which could be put into action quickly. It should also have been able to do without the lengthy work necessary for the firm concrete foundation by depending for solid stability on *Radgürtel*, pedrails, used by heavy guns worldwide since their invention by the Italian Capitano Bonaventura in 1904. These consisted of a dozen or more little rectangular platforms, fastened to the rim of the gun wheels so that they could turn on a horizontal shaft. These greatly enlarged the area on which the gun wheels were standing, distributed the weight onto a larger surface and prevented their sinking into the ground when this was softened by rain. Beneath these wheels were to be placed *Rohrmatten*, carpets of iron tubes, transported in the form of rolls, and onto them a firm but also transportable iron foundation.

But of course the development did not happen this way. All armies of the world work in more or less the same way. So the new design came about because of rivalry between different branches of the service. Here is the slanderous version: the days when artillery and engineers had shared the distinction of forming the *Genietruppe*, the ingenious troops, who depended not only on brawn but also on brain when doing their duty, with one calculating the trajectory of his mortar bomb and the other the necessary dimensions of a wooden beam for the load on a bridge, were long gone. After the reforms of Scharnhorst during the Napoleonic wars, artillery had become the first branch in which officers who were not of noble lineage could make a career. This was remarkable in a time when you would not be accepted in certain regiments, such as the imperial guard, unless you could show by a guaranteed monthly cheque of some hundred Goldmark from Father, that you belonged there and were able to pay for the ridehorses and the champagne in the officers' mess. And artillery had to master scientific high-angle fire now, at least in the *Fußartillerie*. But the poor engineers received no glory, and especially no guns.

They had managed to draw abreast of the other troops a little with the help of one of their men, who was made famous after the 1864 war in Schleswig. When storming the *Düppelner Schanze*, the strong Danish earthworks at Dubbol next to the frontier, Pionier Klinke was carrying his large bag filled with blackpowder (in those days both propellant and explosive) to a palisade obstacle in front of one earthwork. Having lit the fuse he was then hit by several Danish bullets which, though only fired from muzzle-loaders, felled him. So they all went up together – the powder bag, the palisades and poor Klinke. But this sad accident was enough for all the nationalistic German high school teachers, who had been hungering since the Napoleonic wars (labelled *Befreiungskriege*, wars of liberation, in Germany) for a new hero after the famous Major Schill, who fought Napoleon single-handed until his own death at Stralsund, at a time when the King of Prussia thought it wise not to do so.

This was not enough to lift the standing of the engineer corps, however. Having guns of their own would be of more help. Did they not have to clear the way for the infantry to storm the fortifications, blowing up obstacles and then attacking embrasures with *Stangenladungen*, long poles with charges consisting of metal cans filled with dangerous gun cotton, introduced in 1883. Better and safer was the *Granatfüllung* 88, the shell filling of 1888, the German equivalent of the French Melinite and British Lyddit. In the end, however, everyone agreed on the safe Trinitrotoluol, *Füllpulver* 02, filling powder 1902, in Germany; TNT or Trotyl elsewhere.

Now the danger lay in bringing these charges to the obstacles. Creeping up to the muzzles of the defenders with only a woollen coat for protection was no longer state of the art; the charges must be thrown, but certainly not by hand. However, asking for guns of their own would certainly raise the opposition of the artillery, especially the highly influential APK. But they could not really object to something confining itself – at least by name – to the domain of the engineers: the mines. And thus in 1907 the engineers had the men of the *Ingenieurkomitee*, the Committee of Civil Engineers (the branch of the engineers is *Pioniere* in German, and the German *Ingenieur* is a civil engineer who has studied engineering), who were designing the fortifications and counted as members mostly soldiers of the engineers, start work on a special weapon for them. This was named *Minenwerfer*, mine launcher, and the new rival of Krupp's monopoly, Rheinmetall (in 1910 it was still known by its original name of Rheinische Metallwarenfabrik) was working on it.

The engineers got their first launcher in 1910. It was the *schwere Minenwerfer*, with a modest calibre of 25cm (10in) throwing a shell of 97kg (210lb) containing no less than 50kg (110lb) of explosive over a distance of only 600m (1,800ft) at first. This will be discussed in more detail later. Let it serve now to say that the engineers' appetite came with the eating, as a German saying has it, and they went in for a really big launcher.

The situation could not be hidden from the APK for long, and so in 1910 they went to their normal source, Krupp, and asked him to provide a similar mine launcher for them, but bigger and better. Professor Rausenberger, chief designer at Krupp, a man who fathered most of the ideas behind Krupp's big guns in this era, thought this over and declined the idea of a small-bore launcher firing an oversize projectile sticking out of its tube. Instead, in 1911 he proposed scaling up the *schwere Minenwerfer* to 30.5cm Beta-calibre firing a shell with 100kg (220lb) of explosive, the new lower limit set by the APK after trials with Beta shells against the cover of concrete fortifications, protected by an extra layer of rock or soil. In the meantime, even these 100kg of explosive no longer seemed sufficient, so Krupp submitted the designs of two more guns, one of 35cm (14in) firing 150kg of explosive, and the other one of 42cm (16.5in) firing a

*The 42cm* Gamma-Mörser *was too heavy as a bedding gun and took too long to get ready for action. Thus the APK demanded, Professor Rausenberger designed, and Krupp built a lighter version of it on a wheel carriage: the 42cm* kurze Marinekanone *14 L/12 in* Räderlafette, *also named* M-Gerät, *the 'M' standing for* Minenwerfer, *mine launcher. It weighed only 42.6 tons and was transported in five loads on the road, pulled by motor traction.*

M-Gerät *load 1: tube* (right); *and load 2: carriage* (below).

shell filled with 200kg, both at a range of 5–6km (8–9.5 miles). The continuous demands then raised for extended ranges and effect led to Rausenberg confirming that the new Gamma shell could also be fired from this new gun. The APK ordered the first sample of this gun on 15 July 1912. It was called *M-Gerät*, M-equipment, in an attempt by Rausenberger to keep it secret.

The APK was now so convinced of the importance of this weapon that they ordered a second one in February 1913, even before the first one was delivered in December 1913. As was to be expected with such a radical new development of combined high performance and minimum weight, there were teething troubles. The wear on the rifling was frightening at first, but decreased later on (the lands had been too high, something which regulated itself in the course of firing). This gun was taken on extended road trials in February 1914, which showed that the intended motor tractors built by the firm of Podeus at Halberstadt were not sufficient for the 2,100kg (46,000lb) of the gun. Steam locomotives were a lot better, but the APK was not going to accept the accompanying steam clouds on the battlefield. Thus Krupp got together with Daimler to develop a *schwere Zugkraftwagen*, a heavy tractor, and the APK stayed with the heavy Podeus motorploughs.

In March of 1914, the *M-Gerät* was demonstrated to Kaiser Wilhelm II on the shooting range of Kummersdorf south of Berlin. He was very much taken by it. In June, the second *M-Gerät* was delivered, and while it was still undergoing changes deemed necessary by the APK, war broke out. The first battery armed with the two *M-Geräte* was mobilized at Essen, hometown of the Krupp factory, from 5–10 August 1914, and left Essen on the 14th. On 12 September it fired the first round at the Liège fort Pontisse in Belgium.

### German Artillery at the Beginning of World War One

M-Gerät *load 3: the bedding* (left); *load 4: the cradle and the spade* (below); *and load 5: the equipment needed for the mounting* (below left).

The design of the gun had almost nothing in common with the two-years older *Gamma-Gerät*, except for the calibre of 42cm. Thus it was much lighter and more mobile, firing from a wheel carriage and standing on the new type of bedding described above. There the tail of the gun was moved sideways by two steel cables for taking the rough azimuth, the fine one taken by the usual traversing mechanism.

The gun itself was also different. The tube with a reduced L/12 (compared to Gamma) now sported the traditional Krupp horizontal sliding wedge breechblock, but with two innovations. One was the form of the breechblock now being modelled after Rheinmetall's *Schubkurbelverschluss*, shoving crank, invented in 1902, which took only a single movement to open and could be recocked and fired again in case of misfire without having to open the breech, which could be fatal to the gun crew in case the supposed misfire turned out to be a hangfire and the gun fired on open breech. This safe breech had been accepted by the APK for the FK 96 n/A in 1904. The other innovation (for Krupp designs) was that this breechblock no longer moved in a horizontal recess cut into the tube itself, but in a special square breech piece, which was a separate part screwed onto the tube. This also stemmed from both Rheinmetall and the

*German Artillery at the Beginning of World War One*

*Having raised the crane, the special bedding (*Rohrmatten*, tubing) was laid and then the* M-Gerät *was set up by installing the cradle into the carriage* (top), *whereupon the cart of the cradle was retired with the spade still on it* (bottom).

53

*German Artillery at the Beginning of World War One*

*The tube was then inserted into the cradle* (top) *and finally the spade fixed to the trail of the carriage* (bottom).

*Now the 42cm* M-Gerät *was ready for loading (note the return of the sliding wedge-type breechblock) and finally to fire.*

FK 96 n/A. When the tube had been worn out by firing, which happened according to the amount of powder burnt at each shot – for example, in a 10.5cm light field howitzer after 10,000–15,000 rounds (and in a 38cm naval gun after about 100!; this is why the *kaltes Pulver*, the cold powder made by the addition of Nitroguanidin, was such a success in World War Two, as well as the bore-cooling *Additivmanschette*, additive sleeve, of today, an invention of Bofors in Sweden, for the 105mm tank guns of German Leopard 1 and US M 60) – only the tube had to be replaced, with the breech piece normally surviving three or four tubes. The first two *M-Geräte* still sported carriage wheels made from wood, and their tractors for the four loads into which the M- broke apart were motorploughs made by Podeus.

These two were followed by another ten during the war, all of them dying under blowtorches at the end of it. Note that only the 42cm *M-Gerät* is to be called *Dicke Bertha* (according to its designer Rausenberger's memories). The German press, intoxicated by the first successes of the *M-Gerät*, made it into a *Wunderwaffe*, a miracle weapon, and the readers were enthusiastic about Big Bertha. But cooler heads from the artillery branch were more realistic in their evaluation. Thus Colonel Karl Justrow, one of the brains of the APK, who had watched the mortar hammer the Belgian forts, formed a more critical opinion than both the press and the Krupp reports, designed maybe to give a favourable impression of their product. According to Justrow in a book written in 1919 (*Die Dicke Bertha und der Krieg*), the effect of the *M-Gerät*'s fire was nothing fantastic, but exactly what was to be expected from such a big gun.

The shells worked first by impact on the target, which made the hardened point of the semiarmour-piercing shell penetrate 1m (3.3ft) into reinforced concrete, or pierce about 30cm (1ft) of armour, and then the delay base-fused blast effect of such a large charge, which in the so-called *Langgranate*, long shell, L/3.6, amounted to 144kg (300lb) of high-explosive filling. This blew up the roofs and walls of the old Belgian brickwork fortifications. On the other hand, a shell hit on the curved side of an oblique armour cupola tended to make it slide off or even break apart, with the explosive filling simply spread around. And detonating on the concrete surface prematurely only resulted in a big flash, while detonating inside a heavy earth cover on these roofs only shifted this soil around. What was a quick success against the Belgian forts, especially at Liège, did not work against the modern French fortifications of Verdun. The Gamma batteries firing the same shells (in the course of the war both Gamma- and M- were also supplied with lighter shells of half the weight, which increased the then all-important range) were also effective at first, until their unwieldiness was revealed.

If time was of no importance, the Gammas were a triumph. Firing 158 rounds in thirty-six hours at the Verdun fort of Manonvillers reduced this to rubble, and the Russian forts also felt its power. The Belgians did not bother much with a clear logical analysis of the downfall of their expensive forts, and all hopes based on them quickly found two culprits: Krupp, who had supplied a few of the armour cupolas (although few of the big French ones which were knocked out), and the gases. Of the latter they referred to two different sorts – those created by the explosion of the charge of the shell, and those created by their own soldiers. Brialmont had designed the new forts in either a triangular or trapezoidal shape; in both instances the central fighting block containing all the guns in their armour cupolas was separated from the surrounding conveniences of kitchen, toilets, and so on, by the ditch running around it on all sides. Under continuous German fire there was an understandable reluctance among the soldiers to risk their life by crossing the open-air ditch to reach the toilets. And it was a hot August and the water pipes and cisterns had both been cracked by the movements of the ground. (Ground does indeed move due to the shockwaves of an explosion. Frenchmen spending their time guiding visitors around the big fortress Schoenebourg, one of the few Maginotline

works attacked in 1940 by the German army now known as *Wehrmacht*, who had then been stationed at this fort, say that when one of the 42cm shells from the single *Gamma-Mörser* that survived World War One hit the ground nearby, the whole concrete corridor would heave, twist and turn, toppling the soldiers inside. And these corridors were tunnels running 30–40m (100–120ft) underground. But that is exactly how deep a 42cm shell went into soil.)

The effect of this calibre, whether fired by a *Gamma-Mörser* or an *M-Gerät*, can still be seen

*The shell of the 42cm* M-Gerät *had an armour-piercing point. Although my six feet height tops it, luckily I do not exceed its 830kg (1,826lb) in weight.*

*German Artillery at the Beginning of World War One*

*Shown here is the effect of the fire of the* M-Gerät *1914 on the Belgian fort of Loncin of the Liège-girdle. Penetrating the concrete cover over the ammunition store it set off propellant powder and shell filling, the blast killing more than 500 defenders, most of whom are still buried beneath the ruins. The wave of pressure then proceeded into the shafts of the gun emplacements, hurling up armour cupolas of 50–100 tons, depending on the calibres, with their guns, almost 100 metres into the air. When falling down again the heavy cupolas turned so that they impacted onto their seat in the surrounding* avant cuirass, *the armour ring. There they still lie after almost a century, like dead turtles on their backs, testimony to the hellish effect of an explosion inside an explosive store or a powder magazine. The German infantry was so happy that they did not have to take this fort by storming it that they lovingly christened the 42cm* M-Geräte, *of which only two had been ready when the war broke out,* Dicke Bertha, *'Fat Bertha', after Krupp's only descendant.*

at Fort Loncin north of Liège, now a national monument for over 300 dead soldiers resting under its ruins, and also an open-air museum of armour cupolas and guns of 1900. The blast of the explosion of the ammo magazine in 1914 blew up the right half of the fort, with the split running neatly along the so-called capital postern running along the centre line and forming a vertical cut through this postern. With the right side gone and the left one remaining, this blast killed most of the 500-plus Belgian soldiers inside and shot the armour cupolas with their guns of 15cm, 17cm and even 21cm calibre (6in, 6.8in and 8.4in) still mounted inside some 50m (150ft) up into the air, where they turned with the heavier thick top of the cupolas now pointing to the ground and fell down again. This was reported by eyewitnesses from the German infantry, which had advanced close to the fort ready to take it after the end of the shelling. All cupolas managed to hit their own open *avant cuirass*, the ring-shaped base of hard cast iron on which the cupolas rested, and today are lying like giant turtles turned over on their backs. And looking at the concrete around these *avant cuirasses* you can see the real reason for the failing and falling of these forts: Belgian workmanship. The concrete is not ferroconcrete, reinforced by steel rods, but simple compressed beton, which has not been poured into a monolithic form, but put down layer over layer like slate. This is the technical truth behind Bertha's overwhelming success at Liège.

*German Artillery at the Beginning of World War One*

The only Gamma- left was not the only mortar of this calibre range to survive in World War Two; about a dozen 42cm Skoda-mortars also survived, having initially been made by the Austrian gunmakers at Brünn in Austria, then Brno in the Czech Republic. In 1918 these were handed over by the victors to the newly formed Czech state, and taken into German service in 1938. Even larger was the calibre of the new mortar, designed by Rheinmetall in 1937, when the German *Heereswaffenamt*, the army weapon office, asked for it. It was ready in 1939. It was a 60cm mortar, selfpropelled on tracks. At 18,000kg (39,600lb) it was the heaviest self-propelled gun in the world. Six were built by Rheinmetall, the whole half-dozen being referred to as the *Karl-Mörser*, Karl mortar, in honour of the man behind their creation, General Professor Dr Ing. Karl Becker,

*Although the power of the 42cm shell was considered to be sufficient, the reach of the short L/12 tube was not. So the carriage of the* M-Gerät *received a longer tube (L/30) of a smaller calibre of 30.5cm. The resulting gun was named the* schwere Kartaune *or* Beta-M-Gerät *and was fielded in 1918. It is shown here ready to fire (right), and with the tube on its cart loaded onto a railway car (above). Both feature one version of camouflage: white spots. The other was the so-called* Buntfarbentarnung, *multicolour camouflage, where handsized spots of different colours – yellow, brown, green – were separated by black lines.*

*German Artillery at the Beginning of World War One*

*Here a* Beta-M-Gerät *is shown in its well-hidden firing position among trees* (right)*, and during the installation of the equilibrators made necessary by the longer heavy tube* (below)*.*

head of the army weapon office. They also received individual names, from Adam and Eva to Freya, Loki, Thor and Ziu, these last four all Germanic gods or goddesses. One of these *Karl* is rumoured to have survived World War Two in Russia, near the end of which war they were downsized in calibre but upsized in range by changing to the 54cm (21.6in) calibre tubes of captured French railway guns.

Today, mortars are of the smoothbore muzzle-loading *Granatwerfer*-type, their NATO calibre ending at 120mm, although bigger ones of 160mm are fielded by Israel and the CIS states.

## SUPER-HEAVY LOW-ANGLE GUNS OF THE *SCHWERSTE FLACHFEUER*

The 30.5cm and 42cm mortars were not the only big guns of the German army at the beginning of the war. For want of far-reaching heavy guns for the army, as in other countries, the navy helped out. This they could easily do, as they were used to coping with the short life of their own guns (remember: tube life expectancy was only about 100–200 rounds) by buying spare guns when work started on a new warship. In addition, the weapons of old warships were still usable when these ships were wrecked, or guns became available that had been destined for ships that were never eventually built. Naval gunmakers – in Germany this meant solely Krupp – had to start work early so that the guns could be installed during the construction of a ship. In World War Two this fact resulted in eight already existing guns of 40.6cm (16in), destined for the ships of the *Bismarck* follow-on, the 60,000-ton H-class (not named after Hitler, as has been written, but a simple alphabetic order (the predecessors *Bismarck* and *Tirpitz* were G-class)), which were stopped. These eight guns were

59

## German Artillery at the Beginning of World War One

then installed in the *Atlantikwall*, five in Norway and three near Calais in the *Batterie Lindemann*, named after the captain of the ill-fated *Bismarck*. (I would love to tell you to go and see these three army-designed concrete bunkers, each supposed to protect a single gun in its *Bettungsschiessgerüst* C/39, bedding firing rack of 1939, its only splinterproof armour gun house, but the *vestigia leonis* are all gone: first blown up in 1945 by the British, and then buried by the French in the 1980s under the debris of the Channel Tunnel.) In World War One this sort of surplus was in 38cm (15in) calibre. This is how it happened.

We have already discussed the leapfrogging of ships' armour and coastal gun calibre. The same leapfrogging of course went on between the armour

*When the war began, German artillery was low on far-reaching guns. The 10cm guns did not reach far enough and the 13cm guns L/35 were too few. A 13cm L/35 is shown here in firing position* (top) *and on the march* (bottom), *with camouflage painting.*

*German Artillery at the Beginning of World War One*

*The lack of far-reaching guns in the artillery had to be compensated by long-range guns from other sources. The first were the fortifications. Here a 15cm gun in* Schirmlafette, *armour-mounting, with the impressive L/39.2 has been mounted in the* Festen, *the modern forts, of Metz. Besides its tremendous reach compared to the other guns, the KiSL was designed to be transported by rail or road to other firing positions. It is shown here being moved by steam traction on the road* (right) ...

... *and by railway* (below) ...

... *and by* Lastenverteiler *on the road.*

61

and the guns of the warships, and even between guns and guns. Thus naval guns developed both in calibre and range (and weight) upwards and upwards. The British navy had been leading traditionally, and sized their rifled muzzle-loaders up to 18in, which 100-ton gun they wisely never installed on their own warships (only two each in coastal fortifications on Malta and Gibraltar) but sold to the Italians, who were already dreaming of turning the Mediterranean Sea into the *mare nostrum*, our sea, of Roman glory once again. The British were also very close to selling their own outdated rifled muzzle-loading guns manufactured by Armstrong to the new German navy too. This almost happened in 1864, when the first attempt of knitting together the German nations after the destruction of the *Heilige Römische Reich Deutscher Nation*, the Holy Roman Empire of German Nation, in 1806 by Napoleon I, had resulted in a loose organization, the *Deutsche Bund*, the German union. This also needed a fleet, a fact revealed by the wars with Denmark, and warships needed guns.

The APK, already in existence then (it was founded by the Prussian reformer General Scharnhorst in 1808), considered itself equal to the task of designing not only the land guns but also naval guns. Since the calibre of Prussian artillery was limited by habit then to 48- and 64-pounders, they designed a new 72-pounder which was built by Krupp according to their specifications, test fired in 1865 and blew apart, which was declared by the APK to be the fault of Krupp's cast steel, and by Alfred Krupp to be the fault of the APK's design. More test firing in 1866 showed the power of the 72-pounder to be insufficient for punching through the armour put up as a target. So the gun was upsized to a 96-pounder and tested again in March 1868. This test also included a naval gun by the British competitor Armstrong. This time the Krupp gun was lacking both in muzzle velocity and penetration compared to the Armstrong gun. Krupp laid the blame on the old German *Körnerpulver*, grain powder, he had been forced to use by the APK, whereas the British were using the new

*However the 15cm KiSL was moved, it arrived at a new firing position and was installed there. The photo shows peacetime manoeuvres (note the old type artillery helmet and orderly stacked rifles).*

*Other long-range guns came from the navy, starting with the 15cm* Schnelladekanonen *(SLK), rapid-loading guns, ready to fire from a bedding for greater azimuth* (top) *or transported in a* Lastenverteilergerät (bottom).

## German Artillery at the Beginning of World War One

*Even more versatile was the 17cm SLK L/40, on a wheel carriage (below) or on a barge on the rivers of Flanders (inset).*

*Prismatische Pulver*, prismatic powder, an invention by the Russians, who were cooperating with Krupp a great deal then. The same happened on 2 July at the next test firing, with Krupp's shot hardly penetrating the target; the light Palliser shot used by Armstrong had no such trouble.

The *Kriegsmarine*, the war navy, as the navy of the *Deutsche Bund* called itself, was fed up with being tied to the apron strings of the land force APK and with losing time in raising a fleet against the present enemy Denmark (it was just after the war for Schleswig-Holstein), and demanded the quick acquisition of forty-one heavy British guns for its first three armoured ships. For Krupp this meant that all further guns for the *Kriegsmarine* may be Armstrong's old rifled muzzle-loaders, which seemed unacceptable for Germany and especially for Krupp. He fought to get a delay on the order for the Armstrong guns, and for another chance to demonstrate the ability of his guns with his own prismatic powder and steel shot. This test firing took place in August 1868 and was a complete success for Krupp. The target was almost completely shattered by the first hit of a steel shot fired from the Krupp 24cm gun, and the following continuous firing of both competing guns ended with the first fissure after 138 rounds and complete ruin after 300 for the Armstrong. The Krupp tube was still in working order after 676 rounds fired. Thus Germany did not buy Armstrong guns for her warships, which was probably just as well, for I do not believe deliveries would have continued after 1914.

In the era of the rifled breech-loaders and smokeless powder needing longer tubes in order to burn correctly, calibres waxed once more. The 8in calibre of British guns was followed in Germany by 21cm, then 10in by 24cm; and 11in by 28cm. The Royal Navy would always go one step ahead and the other nations would follow suit. Germany stopped at 28cm. They found this sufficient for a long time, as they considered themselves to be a continental nation with only a small shelf sea – the *Deutsche Bucht*, the German Bay – to defend, and the weather there was mostly foggy which prohibited long-distance ranging and therefore the use of far-reaching ship guns. In addition, more of the lighter 28cm guns could be mounted on a ship, and their shells, weighing only about 300kg (630lb), were easier to handle than the heavier ones of bigger calibre. This gave an edge in rate of fire and firepower to the German ships, so the *Kaiserliche Marine* under Admiral Tirpitz believed.

In the end Germany and the Kaiser looked for colonies, being the last country in Europe to do so – even Belgium, Portugal and Holland had some; only Switzerland was smart enough to stay away from this long-term source of trouble. But this meant long-distance sailing, and if need be firing there under clear blue skies, which meant target acquisition on a much longer range than in the

*The 17cm SLK L/40 proved too heavy for the planned traction by horses, so it was put onto railway cars and fired from there in different mountings, rolling back on rollers* (above) *or with an azimuth of 30 degrees* (below).

*Deutsche Bucht.* Thus Germany joined the club of leapfroggers and upgunned her new warships to 12in (30.5cm), and had just finished building the first two ships of the *Bayern-Klasse*, the class of Bavaria, consisting of the name of the patron, *Bayern,* and her sister, *Baden,* both armed with the 15in (38cm) calibre which the British Admiral Fisher had been propagating, when the war began. For the next two ships of this class, the big 38cm guns of their heavy artillery were also finished, as were some spare guns. These weapons were not needed by the *Kaiserliche Marine* right now, but very much so by the Heer, which was lacking in really heavy flat trajectory guns for the job of reaching deep into the enemies' Hinterland to harass troops, depots, railway junctions and other important wartime places. So the Marine kindly passed on all the guns, both old and new, which it could do without, to the Heer.

The army especially loved the heavies. One 38cm gun was positioned as close as possible and as necessary – no need to move within the range of the enemies' field artillery with such guns – in front of the French fortifications along the border with Germany: Belfort and Verdun. The pits for the gun emplacements were dug quickly, but how to mount the guns? The armoured ship turret was not suitable in most cases, so what carriage to use? The answer came from Krupp, of course. After all, they had had to solve the same problem when firing huge ship guns on their range (since 1877 at Meppen in the heather of northern Germany). The long tubes of high-power guns burning slow smokeless powder could no longer be hung up on trunnions in the centre of gravity; the breech end would have dug deep into the ground on recoil of the elevated guns. So the trunnions were moved to the rear and equilibration restored by means of equilibrators. These lifted up the muzzle end of the field gun either by powerful springs or the pressure of compressed air or nitrogen.

On naval guns this was achieved more simply: huge iron plates bolted onto the upper part of the breech end as counterweights served the same purpose. And this was even overdone on purpose, so that the ships' guns were breech-heavy, with the muzzle swinging up by itself. Thus the Krupp designers mounted the big guns' trunnions in a simple framework and tied the muzzle, which was reaching for the skies, down with a thick steel cable connected on the bottom end to a long screw, which was fixed to a nut on the ground. Now the elevation could be achieved simply by turning the bolt up or down in the nut. This had been working fine for decades at Meppen, and worked just as well in the first gun positions of the super-heavy *Flachfeuer*. The only work left was to design a new mechanical time fuse capable of running longer than one minute, the time limit of the present ones, for firing shrapnels at this long distance with a long time of flight. But this too was solved. From 1916 onwards their mountings were changed for railway carriages.

Besides these new 38cm guns, in World War One the super-heavy *Flachfeuer* also sported the following ex-naval artillery, fighting fixed on the coasts or mobile on land. They were named either after the shipclass from which they came, or after other criteria, such as the 38cm Max, named after Admiral Max Rogge, who was in charge of heavy ex-naval artillery.

- 15cm (6in) SLK (*Schnelladekanone*, quick-loading gun, a term used for the new guns after about 1890, which in loading terminated the charges of their *Vorkartusche,* containing about half of the propellant in a silk bag, with a *Hauptkartusche*, which consisted of more powder sticking out of a brass cartridge case also holding the primer. The older types of guns using only bag charges remained as *Kanone,* cannon, calibre length L/45 named *Nathan Ernst* and *Nathan Emil*, with a range of 22.7km (14.2 miles);
- 17cm (6.8in) SK L/40 *Samuel*, with a range of 24km (15 miles);
- 21cm (8in) SK L/40 and L/45 *Peter Adalbert*, with a range of 25.5km and 26.4km (15.9 miles and 16.5 miles);
- 24cm (9.6in) SK L/40 *Theodor Karl*, with a range of 26.6km (16.6 miles);
- 24cm (9.6in) SK L/30 *Theodor Otto*, with a range of 18.7km (11.7 miles);

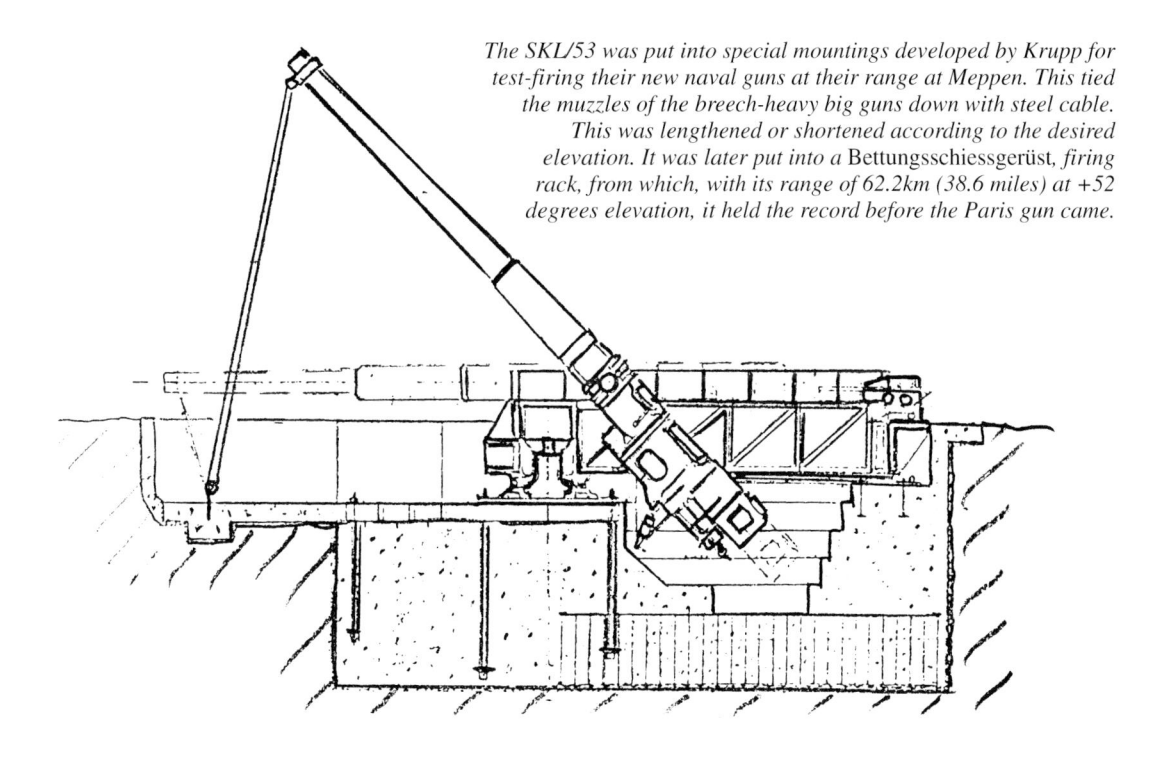

## German Artillery at the Beginning of World War One

*The SKL/53 was put into special mountings developed by Krupp for test-firing their new naval guns at their range at Meppen. This tied the muzzles of the breech-heavy big guns down with steel cable. This was lengthened or shortened according to the desired elevation. It was later put into a* Bettungsschiessgerüst, *firing rack, from which, with its range of 62.2km (38.6 miles) at +52 degrees elevation, it held the record before the Paris gun came.*

- 28cm (11.2in) SK L/40 *Bruno*, with a range of 27.7km (17.5 miles);
- 28cm (11.2in) K (*Kanone*) L/40 *Kurfürst*, with a range of 25.9km (16.2 miles);
- 35.5cm (14.2in) L/52.5 *König August*, with a range of 62.2km at 52 degrees elevation of the land carriage (38.8 miles);
- 35/38cm (38cm tube in 35cm carriage) (15in) L/45 *König Luitpold*, with a range of 48.2km (30.1 miles);
- 38cm (15in) SK L/45 *Max*, with a range of 47.5km (29.7 miles) at 50 degrees elevation. Altogether, thirty-three different emplacements for this artillery are known, most of them in the region south of Strasbourg between Colmar and Neu-Breisach in the Alsace. These important weapons did not fire under the orders of the normal commander of artillery at divisional level. They were reserved for the level of *Armee-Ober-Kommando* (AOK), the high command of an army, the group formed out of several corps. Some were even under the reservation of the *Oberste Heeresleitung* (OHL), the supreme Army Command, which wielded the power that the Kaiser as supreme commander had delegated to the chief of general staff of the field army.

Thus ends the first discussion of super-heavy guns. We shall encounter them again, either in their role as coastal guns or railway artillery, or both. In World War Two they served Germany as railway artillery, also constructed from old and surplus naval guns. After World War Two there was a short revival in the form of the US 280mm self-propelled gun Atomic Annie, styled after the model of the 28cm K 5 (E), of which two had been captured by US troops near Anzio in Italy 1944, and therefore called Anzio Annie. But Atomic Annie no longer depended on railroads; it was free to move along the streets, hung up between two cabs looking like oversized forklifts. Now NATO does not sport anything over 155mm calibre guns on the land.

## Super-Heavy Artillery

| Gun model | Calibre (in) | Weight empl. (lb) | Tube length (in) | Shell weight (lb) | Muzzle Velocity (ft/sec) | Max. range (ft) | Elevation/ azimuth (degr.) | Remarks |
|---|---|---|---|---|---|---|---|---|
| 28cm Mörser L/12 | 11.2 | 35,700 | 136 | 598.5 | 1,038 | 29,100 | +65/10 | (1) Krupp |
| 28cm Mörser L/14 | 11.2 | 32,690 | 158 | 598.5 | n/a | 29,100 | +65/12 | (1) Krupp |
| 30.5cm KüstenMrs Beta | 12 | 41,370 | 104.8 | 861/69 | 930/1,008 | 26,400 | +60/60 | (2) Krupp |
| 30.5cm KüstMrs Beta 09 | 12 | 95,130 | 195.2 | 861/693 | 1,185/1,254 | 35,700 | +67/40 | (2) Krupp |
| 30.5cm H i.R.L.L/17 | 12 | 51,450 | 207.2 | 693 | 1,200 | 35,100 | +75/10 | (1) Krupp |
| 30.5cm H i.R.L.L/30 | 12 | 98,700 | 366 | 693 | 1,800 | 49,500 | 64/20 | (3) Krupp |
| 42cm Mrs Gamma | 16.8 | 315,000 | 268.8 | 1,953/ 1,680 | 1,200 | 42,600 | +66/45 | (4)(5) Krupp |
| 42cm M-Gerät i.R.L. | 16.8 | 89,460 | 200 | 1,680 | 999 | 27,900 | +65/20 | (6) Krupp |

*Remarks:* (1) one gun only; (2) AP-/HE-shell; (3) tube on carriage of *M-Gerät*, also called *schwere Kartaune*, after a heavy old gun; (4) heavy/new shell; (5) transported on ten rail cars, fired from bedding; (6) the one and only *Dicke Bertha*, with armour shield, transported in five loads on road with steam-, later motor-traction, fired from the wheels resting on a *Rohrmatte*.

n/a = data not available

## HIGH-ANGLE WEAPONS OF THE ENGINEERS: THE *MINENWERFER*

We have already encountered the *Minenwerfer*, the mine launchers, in connection with the 42cm *M-Gerät*, the one and only *Dicke Bertha*. The slanderous version of their origin has been told. Here is the other version.

It was another lesson of the siege of Port Arthur in the Russian–Japanese war of 1904–5 that the fire of heavy guns alone was not sufficient to take fortifications, because not all obstacles could be cleared in this way. It also took close-quarter weapons capable of dropping a heavy explosive charge exactly onto a yet undestroyed target a few hundred metres away. Targets such as these included barbed wire obstacles, which the artillery found troublesome to clear. Starting in 1907, the *Ingenieurkomitee*, the Engineer Committee, worked on this. They were responsible for the engineers and everything connected with them, including their equipment. Another of their tasks was the design and building of fortifications and working out regulations for fieldworks. Their industrial counterpart was Rheinmetall, the new firm in competition with Krupp, which had already bested the hitherto monopolistic old master. The best known case was that of the new 7.7cm *Feldkanone* 96 n/A, which had been created by modernizing its predecessor, the old FK 96. Under its dynamic leader Ehrhardt, Rheinmetall had introduced long recoil on guns after the Haussner patents, and also sold such new guns to foreign countries.

Since the APK of the artillery had formed close contacts with Krupp, the engineers found a new partner in Rheinmetall for their rival project of a launcher. They planned this in three calibre ranges: heavy, middle and light. The heavy one was started first and introduced in 1910 as *schwerer Minenwerfer* (sMW), heavy mine launcher. In spite of its remarkable calibre of 25cm (10in), it looked like a downscaled mortar because of its size; this was exactly what it was: a downscaled mortar with a rifled tube, muzzle-loaded and possessing both the recoil mechanism and laying mechanism of a gun.

## German Artillery at the Beginning of World War One

The effect of its 97kg (215lb) pre-engraved shell with its explosive filling of 50kg (110lb), was equal to that of a mortar of 28cm (11in) or 30.5cm (12in), which weighed more than ten times as much compared to the sMW's 955kg (2,100lb) on the march and 660kg (1,450lb) emplaced. It makes one wonder whether the expense of the super-heavy mortars such as Gamma- and M- had been justified, as even they were not able to deal with the modern French forts of Verdun.

When the war started, forty-four of these sMWs had been delivered. Like the other heavies of 42cm, they had been kept secret and thus came as a very unpleasant surprise for the garrisons of the Belgian fortifications at Liège and Namur, and also for the French fortress Maubeuge and the positions on the much embattled *Hartmannsweilerkopf*, where the French *Felsenburg* surrendered after seventeen rounds. In 1913 the secret had leaked into French ears, after all. Thus Supreme Commander General Joffre knew about the *Minenwerfer*, and in October demanded such a weapon from the *direction de l'artillerie*, but in vain. Trench mortars were then as little in favour in France as they were in Britain. There Sir Wilfrid Stokes, Managing Director of Ronsomes & Rapier at Ipswich, a factory of cranes and sluices, had trouble convincing the Ordnance Board of the merits of his first trench gun. In the

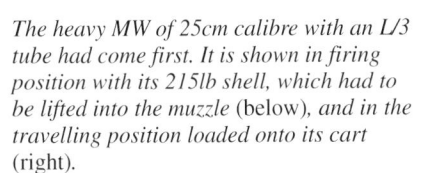

*The heavy MW of 25cm calibre with an L/3 tube had come first. It is shown in firing position with its 215lb shell, which had to be lifted into the muzzle* (below), *and in the travelling position loaded onto its cart* (right).

## German Artillery at the Beginning of World War One

*The sMW also got a longer tube L/5 in 1916 to increase its range* (left). *It is also shown in travelling position* (right).

*A series of 25cm sMW a/A are test fired on the Rheinmetall range at Unterlüss for final inspection. They show clearly the difference between a hangfire, where the sequential firing of the shot is late in developing (MW nos 5 and 7 from the left), and misfire, where the round stays unfired, also shown by the tube not recoiling and staying in forward position (MW no 2 from the left).*

end, he was only successful because Minister of Ammunition Lloyd George saw it demonstrated in June 1915 and ordered 1,000, paying for them with money donated by an Indian maharaja.

The next mine launcher introduced was the *mittlere Minenwerfer* (mMW), the middle one, which reached the troops in 1913 and had a calibre of

17cm (6.8in). It was constructed and looked very much like its heavier brother; in fact, when I have to decide which one is standing before me I have to use a measuring tape on the muzzle. In 1914 the engineers could field 116 of them, and these were successful in the battles around the Antwerpen fortress and in the east. Both were fired sensibly by a long wire and electric primers.

The last of this family, the *leichte Minenwerfer* (lMW), the light one, was finished only in the form of a prototype when the war started. But it was to become the most important of the three. The first version fired from a square bedding. This MW, later called the lMW a/A (*alter Art*, old type) could either be carried by two men with the help of poles or, after two wheels of 75cm (30in) diameter had been stuck onto the axles of its bedding, drawn by hand. In 1916 the next model came to the fore: the lMW n/A (*neuer Art*, new type). The front of the bedding was now shaped like a semicircle and it was able to traverse 360 degrees. Maximum range had been extended from 1,050m (3,150ft) to 1,300m (3,900ft). Later, it also received a tail which enabled it to fire in the flat trajectory mode. This light model was also a rifled muzzle-loader with recoil and laying mechanism, but contrary to the electric firing of the heavy and medium models, this was fired by percussion with a *Wiederspannabzug*, a recocking trigger, and a long rope. It too was loaded from the muzzle, by dropping the shell into the steep tube when firing in the high-angle mode, and by pushing it in with a rammer when the tube had been tilted into the almost horizontal position for the flat trajectory low-angle mode.

(Above) *The medium mine launcher had a calibre of 17cm. Like its brothers it was built like an artillery piece, with recoil brake and recuperator. This is the first model, later called a/A, with L/3.8, showing the electric firing equipment and the shell with the lifting handle still fixed to the nose. After loading this would be replaced by the fuse.*

*The new model of 1916 featured a longer tube with L/4.5.*

71

*German Artillery at the Beginning of World War One*

(Above) *The smallest calibre* Minenwerfer *shown here served by its crew.*

(Right) *The smallest calibre of the* Minenwerfer, *mine launcher, with 7.85cm, was the latest to arrive. Here it is shown ready to fire from the older square bedding.*

The 7.85cm (3.14in) tube of L/5.2 was the same in both old and new models, as was the shell of 4.6kg (9.6lb) with 0.56kg (1.2lb) of explosive. This filling was no longer the shockproof *Füllpulver* 02, as TNT was called by the German forces. The production of this was reserved for the shells of the artillery guns, with their higher loads of shock on firing. The *Minenwerfer* were considered to be less touchy in this field due to their lower muzzle velocity of 56m/s (170ft/s) for the heavy and 63m/s (190ft/s) for the medium MW, in both cases with the

*The new model of the* Minenwerfer *featured a round bedding and a longer tube L/5.2* (right). *It was also heavier, especially with the trail fixed* (below).

*The new model shown in the high-angle mode* (below) *and the low-angle flat trajectory mode* (right), *in both instances with trail stuck on.*

*The crew of four had to pull the* Minenwerfer.

lowest charge. The new light model n/A fired at 77m/s (230ft/s) with the lowest and 121m/s (360ft/s) with the highest charge, ranging between 160–1,300m (480–3,900ft). Guns were sending their shells at 325m/s (975ft/s) for the 15cm sFH 02, and 465m/s (1,400ft/s) for the standard FK 96 n/A., with the ex-naval guns topping this by far, the 38cm (15in) guns, for example, reaching over 850m/s (2,600ft/s). And then there was a certain 21cm gun at double this speed, with an incredible 1,533m/s (4,600ft/s). This weapon will be discussed later.

Thus the *Minenwerfer* ammunition was filled with anything on the market that was likely to explode and was used in mining. The names given to them by the manufacturers were legion: Ammonal, Ammonit, Donarit, Roburit, Westfalit – most of them being ammonium nitrate-carbon explosives. But they were not entirely safe. In spite of the low firing shock, they tended to go off in the tube, splitting it and causing casualties among the crew. Even when firing under the relatively safe conditions of industrial testing and trials, when no enemy was present, they blew up, killing Director Karl Völler, the head designer of Rheinmetall, this way in 1916. Because of these dangerous accidents the troops disliked it for a time, until the teething problems of *Minenwerfer* ammunition had been overcome.

The shells also had two other peculiarities: the pre-engraved driving band, necessary in a rifled muzzle-loader (and still used today with the rifled US mortar of 106mm) and the propellant. This consisted of disks of powder inserted into the base of the shell and held there by a perforated cover, which also held the primer in its centre, electric with heavy and medium, percussion with the light model. Half the fuses were percussion fuses which worked independently of the angle of impact, even if flat, and combined mechanical time and superquick fuses. Initially the shells were made from tubing of Siemens-Martin steel, with forged bottoms screwed in. Later a lack of this material forced a change to pressed Thomas steel. Problems with the manufacture of these also led to fissures and to premature detonations. The next problem was the lack of brass for the driving bands, a result of Germany's eternal hunger for raw materials in every war. Iron solved this problem, both for the driving bands and the small parts of the mechanical fuses.

Things took a turn for the better in June 1917 when the APK took over the *Minenwerfer* section of the *Ingenieurkomitee* (which might have seen this differently). We already know that the shell of a heavy MW was as effective as one of a 42cm *M-Gerät*, with shell and M- each costing ten times as much as the MW. Almost the same held for the light MW. This cost only one-seventh of its counterpart, the 7.7cm FK 96 n/A, which also needed six horses for transportation compared to the single one for the lMW. It could also be moved by four soldiers on the battlefield, who unlike the horses would take cover from enemy fire, giving it a higher rate of survival than the light FK used for fire support of the infantry. The ammunition was a lot cheaper too, since it did not need the precious brass cartridge case. Another advantage of the MW was that it could be manufactured by smaller factories with no experience in gunmaking. These could produce about five times as many MW as FK with the same effort, because of the lower weight and simpler design of the MW. Another advantage was that the tube life of the MW was superior to that of artillery, needing less regular replacement.

The OHL, Supreme Army Command, recognizing in 1916 that this enabled Germany to equal the higher artillery power of the Allies, increased production of the light MW about a hundred times by bringing in additional manufacturers, so that in 1918 the number of MW had increased from the forty-four medium and 116 heavy to 1,234 heavy, 2,361 medium and an additional 12,329 light MW. Even a super-heavy 38cm (15in) was developed: the *sehr schwere Minenwerfer* (ssMW*)*, the very heavy MW.

But all of these would not have been sufficient even if they had already been in existence when the *Stellungskrieg*, the war of positions, began in the middle of September 1914. This created such a tremendous demand for *Minenwerfer* that the industry was unable to cover it. The engineers, used to improvising, were the first to help themselves by developing smoothbore muzzle-loaders in the form of *Erdmörser*, earth mortars, like barrels dug into the ground, reviving the old *fugasses* of medieval times; later with steel tubes fixed rigid on simple mountings without any recoil mechanism.

Then at last German industry took over. Countless manufacturers went into the urgent business of manufacturing makeshift launchers. Looking at the list of names concerned with this one gets the impression that anyone in Germany who could lift a pencil took to designing a *Werfer*, and if he could also wield a file he started building too. The names included Albrecht-, Ehrhardt-, Genter- (also designed by Captain Magener and built in the Pionierpark of the 4th Army at Gent), Heidenheim-, Iko-, a design of the *Ingenieurkomitee* built by the Zawatzki-Werke, Lanz-, Magener-, Mauser- and Voith-Werfer, all of which fired their mines in the usual way. Mechanical means such as compressed springs really threw them from the Bosch-Werfer, while compressed air pushed them from the *Pressgas-MW* of Ehrhardt and Sehmer in Saarbrücken, who built a range of these smoothbores in 10.5cm (4in) and 15cm (6in) calibre for fin-stabilized shells, which were fired almost without noise and ranged up to 800m (0.5 miles). The Marinekorps in Flanders even had them in 26cm (10.4in) calibre. Production was speeded by omitting rifling in both heavy and medium MW. The 24cm (9.6in) *Flügelminenwerfer*, launcher for winged mines, also a smoothbore launcher for fin-stabilized shells, was the German answer to the notorious French 240mm model 240 L, which penetrated deep enough and carried enough explosive to smash dugouts 10m (30ft) deep underground, and was designed by *Iko* and also by Albrecht.

Later, with the rising supremacy of the light MW, there appeared smaller models such as the *kleine Granatenwerfer* 16, little shell launcher of 1916, which for the first time revealed the name that the *Minenwerfer* would later bear in the German *Wehrmacht* (today it is called a mortar in the German *Bundeswehr*, due to the influence of the US in NATO). The *kleine Granatenwerfer* 16 caused problems of a different nature. It had come from the Austrian army, which had introduced the design by a Hungarian named Vécer, who ran a seminary, also being a priest himself. The Austrians innocently called this handy weapon *Priesterwerfer*, priest launcher, and liked it both for the

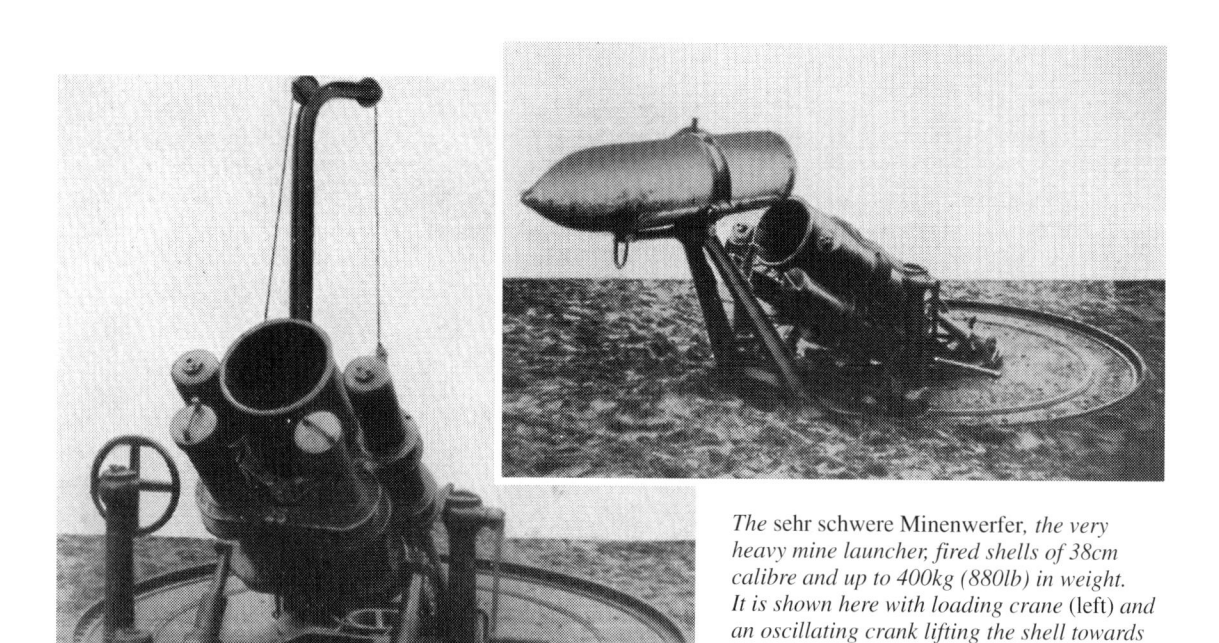

*The* sehr schwere Minenwerfer, *the very heavy mine launcher, fired shells of 38cm calibre and up to 400kg (880lb) in weight. It is shown here with loading crane* (left) *and an oscillating crank lifting the shell towards the muzzle* (above).

*To the many makeshift MW belong the* Ladungswerfer Magener (right) *and the* schwere Albrecht-Flügelminenwerfer (below), *which launched a fin-stabilized shell.*

# German Artillery at the Beginning of World War One

(Right) *The 24cm* Flügelminenwerfer *L/5.25 of 1917 fired fin-stabilized shells from its smoothbore tube, without needing any recoil mechanism.*

(Below) *Another provisional MW throwing strange-looking shells.*

*Other interim systems used compressed air to launch the mines. They fired from square beddings* (below left) *or from ring-shaped ones with integrated wheels* (below).

77

*In the end, all of them finished up on the other side, as shown here on a French postcard.*

*This small spigot mortar was developed by a Hungarian priest for the Austrian army, which named it* Priesterwerfer, *priest launcher, after the inventor. When the German army adopted this successful weapon in 1916, they changed the name to* Granatenwerfer *1916; maybe they feared too literal a misunderstanding of the original name.*

performance of the 2kg (4.4lb) shell and the 200–500m (600–1,500ft) range, combined with its low weight of only 40kg (88lb). In the German army each infantry company was supposed to receive two of these *Granatenwerfer*. But this launcher, built by Stock & Co. in Berlin-Marienfelde among others, kept raising the question: if a *Minenwerfer* throws *Minen* and a *Flügelminenwerfer* throws *Flügelminen*, obviously a *Priesterwerfer* should throw *Priester*, priests. In order not to confuse the infantryman or engineer loading this weapon, the designation was changed to *Granatenwerfer*. The other problem it caused was the fact that it hurt only the receivers of the small fin-stabilized shells: arriving with a whirring noise caused by the sheet steel fins – the French therefore named it *tourterelle*, turtledove – at a low final velocity it hardly penetrated the ground before its super-quick fuse made it explode. Thus most of the fragments went horizontally, hitting targets. The French claimed these wounded more than the light MW.

## German Artillery at the Beginning of World War One

This weapon received another task. The heavy and medium MW were used as high-angle artillery firing with gun director and panoramic periscope. At first they were emplaced into the trenches, but as the infantry did not much like the enemy fire they attracted there, they moved out of the trenches to the rear, often into special dugouts. This demanded both a greater range and a greater azimuth of fire and led to the development of the sMW 16 and the mMW 16, both launchers being equipped with longer tubes and firing larger charges, which increased their ranges to 970m (2,910ft) for the heavy and 1,160m (3,480ft) for the medium model. A turntable underneath both weapons widened their azimuth to 360 degrees.

We shall return to the decisive year of 1916 later on to learn about the other changes, improvements and new gun models it brought. As far as the light MW was concerned they were many. In the time in which his medium and heavy brothers turned into weapons of the war of position, the light MW became the accompanying fire support weapon of infantry, especially after having a tail added to its carriage, which enabled it to flat trajectory fire. Thus it played an important role during the German offensives of spring 1918, when it cleaned out machine gun positions and centres of resistance, stopped counterattacks and even tanks, all in the flat trajectory mode. Its one great advantage compared to field guns was its light weight, which enabled it to be moved even in the otherwise impassable battlefield filled with craters. There were no problems caused by lack of communication between the infantry in front and the artillery some kilometres behind, and it could fire over its own troops by going into the high-angle mode, up to the moment at which they overran the enemy lines. Of course, the targets had to be selected carefully, because ammunition supply became a problem in such a tactical situation. The lMW therefore concentrated on targets the artillery could not combat, either because they had been hidden behind a cover before the attack, or because the target acquisition had not seen them due to the distance.

The *Minenwerfer* were kept by the *Reichswehr*, the defenders of the Reich, which was what the German forces were called after 1918; the name changed, but the *Minenwerfer* did not. Each infantry division of the *Reichswehr* was granted three companies of them, with three medium and eight light *Minenwerfer* each. The further development of the heavy model was forbidden. Attention then centred on the Stokes-Brandt type smoothbore mortars because of their low weight and simple construction, and led in the end to the 5cm (2in), 8cm (3.2in), and at the end of World War Two to the 12cm (4.8in) *Granatwerfer*, shell launcher, which became an infantry weapon. Top secret 15cm and even 38cm *Ladungswerfer*, spigot mortars, were also developed, although not fielded.

The engineers of NATO no longer deal with special mortars, but they have improved the dangerous and tedious age-old business of mine warfare. They no longer plant whole minefields like potatoes, draw exact plans to find them later, or risk their lives in clearing them. The engineer of today will drive along the expected battlefield in his armoured engineer vehicle, a sort of APC-carriage with an armoured cab and a series of launchers mounted on it. From this vehicle he fires literally hundreds of *Wurfminen*, thrown mines, against tanks only. These mines will stop the enemy for a given time – a certain number of hours according to the setting of their fuses – after which they either blow themselves up or switch into the safe mode.

Let us close this section on the poor man's artillery with the poorest of them all: the infantryman (hence its name, *Königin der Waffen*, queen of all branches of service) and the rifle grenade on the barrel of his rifle. These were derived from the wish to give to the infantry some means by which it could throw its *Handgranate*, grenade, over a longer distance. Muskets firing small mortar bombs and grenades had been used in the seventeenth and eighteenth centuries. The grenades were thrown by a special troop in the army of the *Alte Fritz*, old Frederic, as the subjects of King Friedrich II lovingly called him behind his back. These soldiers were

*Another poor man's gun was the rifle grenade fired from the infantry man's rifle. Here the* Kugelhandgranate *13, a grenade destined to be thrown by hand, is fired with a special rod.*

therefore called *Grenadiere*, a name which has returned today to the German Infantry.

The grenade suddenly disappeared from the world stage around 1800, then reappeared a century later in the Russian–Japanese war of 1904–5. Soon attempts were made not to throw it but to shoot it, and over a longer distance too. (A strong man could throw the earlier potatomasher *Stielhandgranate* about 60m (180ft); the newer egg-shaped *Eierhandgranate* up to 100m (300ft); and if he used the *Wurfmaschine*, throwing engine, made by world-famous Bosch, he could bring up to three grenades at once, depending on the type, up to 150m (450ft). We have already encountered this 75kg (160lb) engine among the launchers.) One of the examples of the rifle grenade in the German army was the *Kugelhandgranate* 13, ball grenade of 1913, which was fired with the aid of an iron rod (*Schiesstock*) screwed into it, which was stuck into the muzzle of the rifle. It could be described as another makeshift weapon with a lot of drawbacks, such as heavy recoil and wear on the rifle.

Later in the war, the German army introduced a new rifle grenade, the *Gewehrgranate* 16, rifle grenade of 1916, which was based on the design of the French system of Vivien-Bessières. It exchanged the filling (rifle powder for manoeuvre cartridges) for real explosive and employed a cup-shaped grenade launcher stuck onto the rifle barrel, into which the cylindrical grenade was placed so that a central round hole in it was in line with the axis of the bore of the rifle. Then the rifle was loaded with an ordinary ball cartridge and fired. The bullet passed the hole, thereby grazing a primer which ignited a delay pellet which in turn detonated the cap and this the charge. Hopefully for the firer, in the seconds between firing and detonating, the gas of the propellant ejected the grenade from the cup and threw it 30–180m (90–540ft) depending on the elevation of the rifle barrel.

The grenades were important in World War One, as revealed by the numbers used on both sides. Germany alone used up to 300 million and made eight million a month in the summer of 1917 alone. This enabled the troops to do away with their own improvised substitute grenades made from cans filled with powder and nails or bolts and lit by a piece of fuse, not to mention the obviously suicidal makeshift hand grenades used, according to war correspondents, by the Russian defenders in the Russian–Japanese war. Seizing one of the 7.62cm (3in) shells for the Russian mountain gun, they set the fuse on delay, rammed the nose of the fuse hard onto a firm surface to set it off, and then lobbed the shell into their barbed wire, which was crawling with attacking Japanese. (Knowing most German fuses of World War Two to have 0.15–0.2 seconds in the *Verzögerung*, delay position, and modern fuses to have even less (0.005 seconds), I cannot help admiring the speed of these men.)

The real surprise is what happened to rifle grenades. Experiencing a renaissance in World War Two with the antitank version with its shaped charge warhead, they remained in use in the 1950s and

*German Artillery at the Beginning of World War One*

1960s. They then suffered a sudden decline, being phased out by the small-calibre hand-held missile launchers of *LAW*-type (light antitank weapon), built from plastic tubing and holding a rocket with its shaped charge warhead. The other side used to lug around the RPG-7, copying the last version of the German *Panzerfaust*.

Now rifle grenades are returning. The first sign was the M 203, a launcher for 40×46mm cartridges (NATO differentiates cartridges of the same calibre (first number) by the length of their cartridge cases (second number)) fixed to the barrel underside of the US rifle M 16. This was fired by the half-dozen rounds from exotic weapons such as the Jackhammer, which fires six of them from its cylinder magazine. Then came the reworked .50 calibre machine gun M2, converted to fire a beefed up version of the 40×46mm cartridge in bursts, and which was followed by other models of other calibres.

And now the ultimate weapon has (almost) arrived, one which makes every rifleman into a gunner fielding his own piece of artillery. Inspired by the prospect of still unlimited dollars in the US armament business, the German arms manufacturer Heckler & Koch, founded by these two men after 1945 out of the ruins of the once world-famous Mauser at Oberndorf on the Neckar, a business now in British or at the time of reading perhaps already in other hands, has designed the objective individual combat weapon (OICW). This name hides what looks like a lot of plastic wrapped around a two-barrelled rifle, one of them appearing fatter than the other. As a matter of fact, this is not one weapon, but rather two: one an automatic rifle of NATO calibre 5.56×45mm, and the other a semiautomatic 20mm (0.8in) gun firing explosive shells. This combination is surrounded by a lot of electronics, which will calculate the range of a target by day and night, tell this to a computer and set the fuse of the next 20mm cartridge accordingly. You simply aim the optics at the head of the enemy and the computer

**High-Angle Weapons of Engineers and Infantry**

| Gun model | Calibre (in) | Weight Empl. (lb) | Tube Length (in) | Shell Weight (lb) | Muzzle Velocity (ft/sec) | Max. Range (ft) | Elevation/ Azimuth (degr.) | Remarks |
|---|---|---|---|---|---|---|---|---|
| Granatenwerfer 16 | n/a | 84 | 6? | 4.2 | n/a | 750–1,500 | + 80? | (1) Vécer |
| 7.58 cm lMW n/A | 3 | 294 | 16.4 | 9.66 | 231–63 | 480–3,900 | + 75/360 | (2) Rheinm |
| 17 cm mMW | 6.8 | 1,123 | 25.8 | 113 | 189–261 | 3,480 | + 75/20 | (3) Rheinm |
| 24 cm sFlügelMiW | 9.6 | 1,302 | 50.4 | 206 | 255–390 | 4,500–6,600 | + 75/360 | (4) n/a |
| 25 cm sMW | 10 | 1,386 | 30 | 204 | 168–219 | 2,910 | + 75/20 | (3) Rheinm |
| 38 cm ssMW | 15 | 2 tons | n/a | 840/210 | n/a | 1,200–4,500 | n/a | (5) Rh? |

Also used literally countless makeshift rifled and smoothbore MW-mortars like Albrecht-; Ehrhardt-; Genter-; *Iko*-; Lanz-; Voith-MW. Others worked by compressed springs (Bosch) or gas (Ehrhardt & Schmer).

*Remarks:* (1) Austrian model also built by Stock at Berlin, also called 'priest launcher' after the designer's profession, a spigot mortar firing a super-calibre fin-stabilized shell; (2) also firing wih an extra tail in flat trajectory; (3) old-/16-model; (4) old/new model; (5) *sehr schwerer* MW, very heavy MW.

n/a = no data available; Rheinm = Rheinmetall.

will send the shell over his head with the fuse set to detonate exactly at that point. No more wasting thousands of rounds to make a single kill; no need to shovel fired cartridge cases out of your APC after an hour of combat. Every soldier is now both an infantryman and a gunner. Will the eternal peace, which we were promised after each war and after each new weapon, finally break out? Read the above and join me in doubting it.

## MOUNTAIN GUNS

What are mountain guns? In the *Militärlexikon*, the military encyclopaedia, written by Lt Col Frobenius in 1901, a standard authority on military questions of the old empire, under *Gebirgsartillerie*, mountain artillery, is the following definition (my translation): 'Movements of field guns would be difficult in the mountainous terrain of the Alps or even completely impossible. All greater nations therefore have a special mountain artillery, except Germany.' This seems to suggest that this will be a brief chapter. But let us look at various different countries. There are numerous German mountain guns to be seen everywhere, built by Krupp and by Rheinmetall, so who needs and uses them? Obviously countries with a lot of high and mountainous terrain and – we start in the last century – almost no roads in the mountains. Taking an ordinary field gun up into these mountains would involve a lot of strenuous lifting, with the bodies of neither men nor animals being able to exert themselves as they could at sea-level or on a plain.

> At 3,000m (9,000ft) you are grateful to get your breath with only your own body weight; you really do not need the extra burden of a few hundred pounds of a gun, and if it is winter and you have to wrap up warm to keep alive, the gun has to be as light as possible. It is much more comfortable to transport if it can be taken apart into small pieces. The weight should be around 120kg (250lb), so that four men can handle it. And what about the tube? This is heavier and weighs about 300kg. Well, it will need to be cut off until it reaches about 120kg. But this will give us only a short tube, with a lower muzzle velocity (300m/s; 900ft/s) than the field gun (500m/s; 1,500ft/s) and will not fire over a long range. Well, with an elevated gun position ballistics will give us a bit more and this will have to suffice. Can we also make the tube into two or three pieces? Well, it has been done before by the British, with their 3.7in mountain howitzer Mk I. They joined the two halves of the tube using a bayonet joint. Calibre? Small, about 60–70mm (2.4–2.8in), compared to the field guns of 75–77mm (3–3.1in). Effect? What do we care about this. The most important thing is that we can manage to bring the pieces of this light gun up to the top of the mountain, where we can have a wide range of the enemy territory before our eyes. From these mountains we need a large azimuth and therefore a split trail carriage. This is too heavy. A simple carriage will have to do. And you need not ask for an armour shield on the gun for splinter protection as this is also too heavy.

Thus ends the imaginary discussion between a mountain gunner asking for a new gun and the designer, in which most of the problems have been covered. In the end they are all reduced to one problem: weight. When, in about 1870, those countries with a mountainous border separating them from their neighbours – and generally long-time enemies, such as the freshly reunited state of young Italy and the old empire of France, or the ragged Donaumonarchie of multinational Austria and upstarting Italy – investigated the possibility of bringing the new rifled breech-loaders up to the border fortifications, sometimes built long ago, they learned the above in a short time. Of course, arming the fortifications was no problem. When erecting them roads had had to be built to transport rocks, sand, lime, and especially water for the masonry. The finished fort also had to receive its guns, together with powder and shot. Add the furniture and the provisions for the garrison – all were transported upwards, except for the soldiers who had to walk.

Later, cement and armour were needed. These too could be moved up the existing roads by draught horses or by men. But in terrain with only a narrow footpath on which to bring up the guns, it was either carry them or find some other means of doing this. Two different animals competed to serve mankind; which one was favoured depended on the country and the time of the decision. The two possibilities were the horse and the mule, both with their own advantages and drawbacks. The horse was more willing, but not as sure on its hooves as the mule, which is famed for its obstinacy. So the troops had to resign themselves either to watching the horses take a vertical shortcut downhill, with all parts of the gun, or having to kick a mule every step of the way uphill.

Luckily, Germany was not concerned with these problems, opting to follow the Swiss by selling arms to any side able to pay for them. Germany's manufacturer of the time was Krupp, who built and sold exactly 688 mountain guns, of the old pre-Haussner design with recoiling carriages, between the 1870s and the end of the century. Their calibres ranged from 6–8cm (2.4–3.2in), they had whatever Krupp-made breech was in at the time of their manufacture, and at the end of that period were equipped with the *Federsporn*, the shock-absorbing tail. After Krupp had bought up Gruson, the Gruson quick-fire guns in 3.7cm, 5.3cm and 5.7cm (1.5in, 2.1in, and 2.3in) calibre with the new vertical sliding breechblock turned up in the Krupp catalogue, as did the 5cm (2in) *Schnellfeuer-Gebirgskanone* M 1903, rapid-fire mountain gun of 1903, with L/15, which could either be distributed among four ponies or have all of its 194kg (410lb) loaded onto one elephant. Like all other European gunmakers of this time, Krupp also made what were called *Buschkanonen*, colonial guns, and if you read the old advertisements explaining why the small calibre of 3.7cm (1.5in) was sufficient for the natives (the running targets), you would see that political correctness had not been invented yet.

Later Krupp lost his monopoly. This was due to Ehrhardt, the director of the new competitor Rheinmetall, who had the foresight to recognize the possibilities offered by the Haussner invention of long recoil, and acted on this. The new quick-fire guns with long recoil found their way first to other countries, and eventually into the German artillery. Prior to this, even the public had finally realized that the artillery of the Kaiser was armed with both a field gun and a field howitzer which had been superseded, and that much better guns were being sold abroad. This led to a heated political discussion in the *Reichstag*, the German parliament, on 10 December 1903, during which the leading socialist August Bebel pointed out that the new gun was already in existence when the army bought the old ones. General von Einem, then Minister of War, defended his position by saying that 'If I had to choose between the new French 75mm model 1897 and our old gun (FK 96) I would take the latter.'

In spite of this declaration for Krupp's antiquities, Rheinmetall flourished and sold a lot of guns abroad, including mountain guns of their own modern design. They even supplied some to the German army, a dozen forming three batteries of their model 7.5cm *Gebirgskanone* L/17 M 08. But again these went to another country: to the *Schutztruppe*, the small colonial army in Deutsch Ostafrika.

Rheinmetall also made mountain howitzers, which could fire a larger calibre shell of 10.5cm (4in), and this in the high-angle mode needed in the mountains. Both Rheinmetall and Krupp also followed new lines of design, such as guns firing with the tubes in counter-recoil, and with vertical breechblocks opening on recoil and closing on the new cartridge automatically.

Thus the table was well laid for the German army when the war started, even though they did not have a single mountain gun (outside of Africa). They soon discovered a need for them. The fighting in the Vosges, the mountains south of Strasbourg, revealed that the field artillery were unable to follow the infantry up the steep mountains. And the flat trajectory FK were also unable to get to grips with the targets hidden behind high mountains. But this was nothing compared to what happened to the *Deutsche Alpenkorps*, the German alpine corps, when it came to the aid of its Austrian allies fighting in the Alps and Carpathian mountains.

## German Artillery at the Beginning of World War One

*The 7.5cm* Gebirgskanone L/17 *of 1908 by Rheinmetall was used in the German colony of* Südwestafrika. *For firing, its tail could also be shortened.*

(Left) *The only new development in German mountain guns during World War One was this 7.7cm* Gebirgskanone L/17 M 1915 *by Rheinmetall.*

As a beginning, the mountain guns already captured or still on hand would have to do. Both makers had some left. Rheinmetall still had three batteries of six guns each of the 7.5cm *Gebirgskanone* L/16 M 1914, originally ordered by China but not delivered because the German army had laid their hands on them at the outbreak of the war. They were without ammunition, but this was remedied when Rheinmetall converted captured Belgian ammunition. Half the guns went to the German army and the rest to the Austrian allies. Later the Turks received most of them. Krupp had four 7.5cm *Gebirgskanonen* L/14 M 1913, destined for Chile. These were used by the newly formed *Gebirgsartilleriebataillon* Nr 1.

The captured guns were French and fired with their tubes on the counter-recoil, something which the French maybe loved more than the new German owners, who complained about the lack of accuracy of these guns. These were the *Gebirgskanone* 45mm M 1906 and the same calibre *Gebirgskanone* 65mm models by St Chamond. Better liked were the Russian 7.62cm *Gebirgskanone* 1909, French guns made by Schneider. These were altered by Krupp to fire German 7.7cm field gun ammunition. A new development by Rheinmetall resulted in the 7.7cm *Gebirgskanone* L/17 M 1915, which was lightened by using a lot of steel tubes in its construction. These guns also later went to the Turks, when Germany received the even better Austrian mountain guns.

The calibre of 10.5cm (4in) of the light field howitzer seemed more promising. Both Krupp and Rheinmetall had howitzers in this range: Rheinmetall the 10.5cm *Gebirgshaubitze* L/12 M 1912, and Krupp the 10cm *Gebirgshaubitze* L/12. But together they only had three weapons, which were altered to fire the ammunition of the lFH. Krupp changed the design of his gun to the 10.5cm *Gebirgshaubitze* L/12 and built eight batteries of these. Two went to the German army, four to their Bulgarian allies and two to the Turks, who were also on the German side.

In addition, Germany and Austria worked on standardizing their mountain artillery, with the Germans wishing to profit from Austria's greater experience, and introducing the Austrian 7.5cm *Gebirgskanone* made by Skoda, which was later to

84

## German Artillery at the Beginning of World War One

replace the existing German mountain guns. A battery of 10cm M 16 *Gebirgshaubitzen*, also made by Skoda, was then altered to take the German 10.5cm ammunition of the lFH and tested by Germany. They also cooperated in reworking both the 7.7cm FK 96 n/A and the 10.5cm lFH 98/09 for transport in the mountains. This involved breaking the guns into two, and later three loads. The satisfying results led to the APK using this principle on heavier guns too. Thus the following guns were converted for mountain transport: 15cm sFH 02 and sFH 13; 10cm K 04 and K 14. In the end, Rheinmetall built one example of a 10.5cm *Feldkanone* L/35 (*zerlegbar*), 4in field gun (easily dismantled).

In total, German mountain artillery during the war numbered only seventeen batteries of mountain guns and four batteries of mountain howitzers. These formed seven *Abteilungen*, as artillery battalions were then called. Each of these consisted of either a two gun and one howitzer battery, or three gun battery. They normally stayed on the corps level, which sent them to the divisions. Most of them fought in the *Alpenkorps*, but the rest were used on all fronts.

In World War Two the *Wehrmacht* had a whole range of mountain guns in 7.5cm and howitzers of 10.5cm calibre. Today, all NATO nations use the Italian 10.5cm *Gebirgshaubitze*.

(Above) *Experiments to adopt heavier guns for mountain warfare were made with this 10cm* Kanone *04, transported in two loads, with the tube (top) and the carriage (bottom) on their own carts.*

(Right) *The same applied to the 10cm* Kanone *14. Shown here is the carriage with the armour shield on its own wheels.*

*The 15cm sFH 02, shown here with its two loads of tube* (top) *and carriage* (bottom).

### Mountain Artillery

| Gun model | Calibre (in) | Weight Empl. (lb) | Tube Length (in) | Shell Weight (lb) | Muzzle Velocity (ft/sec) | Max. Range (ft) | Elevation/ Azimuth (degr.) | Remarks |
|---|---|---|---|---|---|---|---|---|
| 7.5cm Geb K 08 | 3 | 1,109 | 51 | 11.1 | 900 | 17,250 | +38/5 | (1) (2) Rheinm |
| 7.5cm Geb K 13 | 3 | 1,018 | 42 | 11.1 | 900 | 14,700 | +26/4 | (1) Krupp |
| 7.5cm Geb K 14 | 3 | 1,031 | 48 | 11.1 | 840 | 14,100 | +38/5 | (1) Rheinm |
| 7.7cm Geb K 15 | 3.1 | 1,165 | 52.4 | 14.4 | 930 | 17,700 | +35/5 | (1) Rheinm |
| 10.5cm Geb H 12 | 4.2 | 1,775 | 50.4 | 33.2 | 759 | 14,700 | +75/8 | (3) Krupp |
| 10.5cm Geb H 12 | 4.2 | 1,712 | 50.4 | 33.2 | 768 | 15,600 | +48/6 | (4) Rheinm |
| 7.5cm Geb K 15 (A) | 3 | 1,287 | 46.2 | 13 | 1,050 | 21,000 | +50/7 | (5) Skoda |

Also used in the German colony of East Africa were captured British and Portuguese mountain guns; the latter were the 7.5cm Rheinmetall *Gebirgskanone* M 1905 L/15, sold to Portugal before the war. Two batteries were armed with ex-Russian 7.62cm GebK 1909, a French Schneider gun, rebored by Krupp for 7.7cm cartridges. Also used both 10cm K 04 and K 14, and even 15cm sFH 02 in mountain versions.

*Remarks*: (1) cartridges; (2) only in *Schutztruppe* in Africa; (3) separate loaded; (4) ammunition of the 10.5cm lFH; (5) an Austrian model.

## ANTIAIRCRAFT ARTILLERY

At the beginning of World War One, only Germany could field antiaircraft guns. This may be cause for surprise, especially considering that at this time there appeared hardly any need for these, the enemy aircraft content to observe proceedings but otherwise doing no harm. This development dated back to the painful experiences of the war of 1870–71. Then the Prussian army had surrounded and besieged Paris. This did not keep the French republic government left there (Emperor Napoleon III had been captured with most of his army at Sedan and had been a prisoner at Kassel ever since) from contacting the unoccupied part of France. Minister Leon Gambetta was at the forefront of their resistance, which pinned its hopes on raising new troops to make up for the losses. The first attempts involved sending letters fixed to carrier pigeons; the Prussians sent trained hawks to catch them. It is not reported whether the birds of prey were successful, but the next means of French aerial communication would certainly have been too big for them. Balloons departed the city slowly with a favourable wind towards the unoccupied south. Prussian siege guns could not be brought to bear on them – the carriages permitted only an elevation of up to 10–15 degrees – nor would they have scored any hits with their slow rate of fire. And the Dreyse needle rifle was not up to its Chassepot counterpart in range, nor could it score a mobility kill, as we say today, against the soft skin of a balloon. So the Prussians, annoyed by the French habit of politely waving their folding top hats at them when slowly drifting over the iron ring around the city, decided it was time for a gun that could solve the problem. And for guns there was only one address to look to in Germany: Alfred Krupp, Villa Hügel, Essen/Ruhr.

In a surprisingly short time the answer arrived in the form of five guns of a completely new design: the 3.7cm (1.5in) *Ballon-Kanone*, the balloon gun. This looked a little like an oversized rifle and was mounted on a pivot which rested on a four-wheel car drawn by horses. The idea was not to sit and wait for the balloon to drift out of range slowly, but to follow it across country until it came into range, and then down it. But neither the horses nor the performance of this antiaircraft gun lived up to expectations, especially when the balloons wisely used the cover of darkness and started at night. In the end, artillery fire and hunger took the city, in which a new revolution, another French specialty, then took more lives than the entire war before it.

Aircraft kept developing, with the first Zeppelin airship flying across the Bodensee on 2 July, and Wilbur Wright flying for the first time with a fixed wing motorized plane on 12 December 1903. This became something the military could use too, and so in 1906 the Prussian Ministry of War (Prussia did all the planning for the other German nations, united since 1871 in the *Kaiserreich*) ordered the APK to assess the fitness of the army guns for combatting dirigibles and planes. The APK returned at once to say that of course they needed special new guns for this. And these seemed to exist already, since Rheinmetall, the progressive, uprising gunmaker, had already designed a weapon of this nature and exhibited it at the Berlin automobile exhibition in 1906. It was a 5cm (2in) pivot gun of L/30, which was mounted on a lightly armoured motor car, firing fused shrapnels up to 4.2km (2.6 miles) with a muzzle velocity of 450m/s (1,350ft/s) and shells with a velocity of 572m/s (1,716ft/s). In turn, Krupp topped this with a 6.5cm (2.6in) gun which reached a ceiling of 5.2km (3.25 miles). Not believing in the need for special guns, the War Ministry held their own trial in March 1907, test firing the guns of field artillery FK and lFH, and even the 10cm K of foot artillery, which bore out the earlier verdict of the APK. Still the Ministry held to its thrifty decision, only admitting the need for range finders.

Between 1908 and 1910 a lot of new guns appeared, necessitating a new trial of AA-guns in 1910. This displayed the guns already mentioned, as well as the new ones, such as a 7.5cm (3in) L/35 Krupp gun on wheels, a 7.1cm (2.8in) L/30 gun on a motor car, and another by Rheinmetall of 6.5cm

## German Artillery at the Beginning of World War One

(2.6in) L/35 calibre. Muzzle velocities were higher, but the pivots remained. Now, together with the APK, the ministry of war laid down rules for *Ballon-Abwehrkanonen* (Bak), the antiballoon guns. These demanded:

- the calibre and ammunition of the 7.7cm field gun, the FK 96 n/A
- transport by a field carriage the weight of that of the field guns of riding artillery
- devices for a rapid change in azimuth and elevation
- or mounted on a motor car with a pivot.

This showed that the use of horses was still preferred by the *Generalstab*, the general staff, which was not exactly promoting modern inventions.

In 1911 Krupp exhibited his solution in the form of a 7.7cm Bak L/27, on a carriage like that of the older 6.5cm Bak L/35. Rheinmetall used the carriage of the lFH, mounting in this the tube of the FK 96 n/A, standing on a wheel bedding. This was to become the *Kanone in Haubitzlafette* (KiH), gun in howitzer mounting, which is exactly what it was. Both manufacturers were not satisfied with their own makeshift solutions, however, and so one year later, in 1912, brought out guns of the FK 96 n/A family, pivoting on motor cars, which was what they had already proposed in 1906. But even then the Ministry demanded a better solution. This continued until 1914, so that at the beginning of the war Germany had available six motor Bak of 7.7cm L/27, two wheeled 7.7cm Bak L/27 with pivots, and ten mixed older models of experimental 7.7cm Bak dating back to 1910–1914.

In April 1914, the general staff had already given orders to procure for each of the eight *Armee-Oberkommando*, army high command, four Bak each, but the Ministry dawdled in passing the orders to the industry and they were of course not finished when war broke out. The experimental guns had been used since spring 1913 for training the crews at the gunners' school at Jüterbog near Berlin. Contrary to the field artillery, these motorized platoons had a range finder with a base

(Above) *The very first antiaircraft gun: the Krupp 3.7cm* Ballonabwehrkanone, *antiballoon gun, of 1870. It was designed to stop French balloons leaving and arriving in Paris during the siege.*

*The project of a 7.7cm L/27* Sockelflak, *AA gun on pivot mount, on a light truck, was realized both by Rheinmetall and Krupp in 1912.*

*In 1915 the first makeshift antiaircraft guns made from captured guns were fielded. This ex-French 75mm M 97 was converted by Krupp to fire the 77mm cartridges of the FK 96 n/A. It has been mounted by the crew on boxes on top of a hill to increase elevation.*

of 1.5m (4.5ft) working on the inverted-image-coincidence principle, which worked well up to 4km (1.6 miles). A fuse setting mechanism had also been tried. Since this would have meant exchanging the German mechanical clockwork fuse for a pyrotechnic fuse of the French pattern, this was refused and more gunners provided for fuse setting. Reading this today one must be reminded of the performances of dirigibles and planes at this time. In 1914 this meant flying at 2,000m (6,000ft) with a velocity of about 100km/h (60mph). A 7.7cm gun with L/27 was therefore thought to be sufficient.

With the start of the war, the horsedrawn Bak were emplaced near bridges over the River Rhein at Düsseldorf and Mannheim, at the Zeppelin wharf at Friedrichshafen, and at the dirigible hangar at Metz. The six motorized Bak went to the 4th, 5th, 6th and 8th Army and two to the 7th Army. None of them knew clearly how to come up to their task, and could not have done this anyhow because of their low numbers. The six motorized Bak were emplaced to protect areas and objects important for mobilization. With all the Bak gone, the troops had to defend themselves with rifles and machine guns, sometimes even with a measure of success.

This forced the War Ministry, which in August had finally ordered 100 Bak from the industry, to confiscate all Bak in construction in the gun factories for foreign countries and to buy all experimental Bak. This raised the numbers available in October 1914 to nine motorized and twenty-seven horse-drawn Bak. So the troops knew they had to help themselves.

An attempt was made to put the FK 96 n/A on a pedestal and to use this together with the lFH 98/09 for combatting aircraft. The low muzzle velocity of the latter soon led to the withdrawal of that much-needed weapon from this role.

The next step was the use of captured guns. Belgian, French and Russian field guns were transformed into Bak, which in January 1915 had grown in number to ninety-seven. Also increased was the danger of enemy planes, which by now had started attacking the troops with machine guns and light bombs. Thus the general staff had to ask for more effective antiaircraft guns. They were to be in the calibre range of between 8–10cm (3.2–4in), the latter the calibre of the rare 10cm K. 04 or 14, which proved to be successful in its antiaircraft role, but which was also much needed in the field for long-range work.

Another makeshift solution was found in the casemates of the German fortifications. These had been armed for close defence of the ditches with a copy of the five-barrelled revolving 3.7cm (1.5in) Hotchkiss gun, a weapon built under Hotchkiss licence by Gruson at Magdeburg. This gun had been acquired first by the *Kaiserliche Marine* for defence against torpedo boats, and in 1884 was also installed in the caponniers of the new forts built since 1872 along the German frontiers in the east and west. There they were not as popular for reasons that shall be covered on page 101; suffice it to say that some 500 of these were lying in the depots. Now some were fixed to a rough mount, bolted together close behind the lines and set up for air defence. They found a comrade in calibre already waiting there, which had been donated by the *Marine* to the *Heer*: about 100 of the 3.7cm (1.5in) automatic machine cannon – though with a more powerful cartridge – of Maxim design, which, like his machine gun, Germany had introduced and was also building under Vickers-Maxim licence at Berlin-Spandau, where both the *Königliche Gewehrfabrik*, the Royal rifle manufacturer, and the *Deutsche Waffen und Munitionsfabrik* (DWM), German Arms and Ammunition Factory, of Ludwig Loewe were put to this task. (The site of these factories at Spandau was engraved on the receivers beneath their names, causing some to confuse this with the name of the weapon itself; in certain literature you will often find German machine guns of both wars referred to as 'Spandau').

Another gun retired to the fortifications was the old workhorse of the 9cm (3.6in) field gun C/73 and C/73/91, which had been superseded firstly by the old type 7.7cm FK 96, and later the modern FK 96 n/A. This C/73 was deemed good enough for the troops of the reserve destined to man the *Armierungsstellungen*, the wartime defence positions, which had already been prepared in peacetime: ditches, shelters, concrete gun positions and so on, around the big fortresses such as Strasbourg, Metz, Diedenhofen (now Thionville). For each of these places literally hundreds of C/73 were hoarded in the artillery depots. Now, in spite of their antiquated design, without recoil of the tube, they were put onto all sorts of improvised mountings supplied by private industry under names such as Koebe-Protze, Metz, Gerät Plett, Schaffhausen, Schnetzler-Sockel, Wohlgemuth.

By mid-April 1915 there were 198 Bak, and at this time the first ones converted from captured guns also arrived on the front. The French 75mm M 97 had had her tubes reamed by Krupp to 77mm in order to take German 7.7cm ammunition, and been turned into the 7.7cm Bak L/35. This became one of the standard guns of German air defence, with 400 of them serving until the end of the war. After the old tubes were worn out, new ones of German manufacture replaced them. July 1915 saw 420 Bak in action. Russian guns could not be rebored due to the brittle tubes, so they kept their 7.62cm (3in) calibre, and when the captured ammunition had been used up, new replacements were manufactured in Germany.

By now the enemy planes were attacking both the observation balloons of the artillery and the infantry fighting on the ground. These could not be combated by the present Bak with their sighting and fire control equipment. So guns with a higher rate of fire were needed. The 3.7cm Gruson-Hotchkiss revolving guns, and especially the ex-naval Maxim automatic cannon, took over this task. So many different guns made training in handling each of them a difficult process. Everyone had to find his own solution to the problems involved with antiaircraft fire.

In 1914 the technical branches – APK, *Gewehrprüfungskommission* and *Ingenieurkomitee* (Artillery Testing Commission, Rifle Testing Commission and Engineers' Committee) – had suffered the loss of their offices; they had all been closed down at the beginning of the war. Now it was recognized that they were still needed. So in the spring of 1916 the officers who had served before therein were called back from the front. Raw materials were becoming scarce, particularly alloys needed for the steel of the tubes, copper and also powder. A new office was founded to organize the procuring of these,

*German Artillery at the Beginning of World War One*

*The troops had to help themselves at the beginning of the war, using all sorts of guns in all sorts of mountings. For light AA guns this started with the 3.7cm Revolverkanone of Gruson, which had been resting in the depots, mounted on makeshift firing racks, either stationary (left) or on wheels (above). The revolver gun was hand-operated, and its rate of fire was between 30 and 60 rounds per minute, depending on the stamina of the operator.*

called *Waffen- und Munitions-Beschaffungsamt* (Wumba), Office for Procuring Arms and Ammunition. The new branch, *Flakwaffe*, a combination of *Flugabwehrkanone*, antiaircraft cannon, and *Waffengattung*, branch, later simply abbreviated to Flak, had come of age. In September 1916 the *Oberste Heeresleitung* (OHL), the supreme command of the army, demanded that the Flak:

- be enlarged and technical competence increased
- be put under the command of the AOK, the high army command. This meant that the divisions would lose it.

The Flak also got a general of its own, the Commanding General of Air Defence or *Kogenluft*.

*Next came the ubiquitous 2cm Becker automatic gun, also an aircraft gun. Firing this from the uncomfortable rest must have been torture. There must have been a second gunner to exchange the empty magazine for a full one.*

## German Artillery at the Beginning of World War One

*In 1914 the* Kaiserliche Marine *handed over 100 of their 3.7cm Maxim automatic guns to the army, which used them to defend fortifications and the like against attacks from the air.*

*(Below) Another gift was received later, when the* Marineflieger, *the naval aircorps, turned over their 3.7cm* Luftschiff *Flak, no longer needed because the era of the Zeppelin had been ended by the advent of the incendiary bullet.*

As with other guns and branches, 1916 was the year when the seeds of 1914 and 1915 finally ripened, and the weapon systems demanded finally reached the troops. For the Flak these came in December and consisted of two guns each of *Kraftwagen* (Kw), motorized Flak, of the calibre:

- 8cm (3.2in) from Krupp
- 8.8cm (3.5in) from Krupp
- 8.8cm (3.5in) from Rheinmetall
- 10.5cm (4in) from Krupp
- 10.5cm (4in) from Rheinmetall
- 10.5cm (4in) *Eisenbahnflak*, railway Flak, from Rheinmetall.

*German Artillery at the Beginning of World War One*

*Auf einer beweglichen Drehscheibe ruhendes Ballon-Abwehr-Geschütz.*

*For heavier calibres against higher flying enemy planes, bigger calibres were fielded. At the beginning of the war these were German guns such as the 9cm C/73, but eventually were mostly captured guns such as this French 75mm M 97* (left), *raised onto a turntable …*

These modern experimental weapons were all mounted on pedestals on motor cars. Krupp later changed his 10.5cm (4in) mobile Flak into one for fixed positions, correctly arguing that these were too heavy for use in the front line. At the end of 1917, of these guns 46 × 8cm, 52 × 8.8cm and 6 × 10.5cm (railway and fixed) were available.

At the end of the war this number had increased to 78 × 8cm, 160 × 8.8cm, 6 ×

*… or this Russian 7.62cm field gun model 00 or 02* (above).

*The target acquisitioning started with the observer watching the sky for enemy planes, which then were ranged with the range finder. This gave both elevation and azimuth to the battery. This rangefinder with a base of 2m was fielded in 1916 and worked up to 6,000m.*

10.5cm railway and 38 × 10.5cm fixed Flak. This number still proved insufficient, so Krupp altered captured Italian guns of the 7.5cm (3in) FK 06 and 11 into Flak for fire barrages. Despite the fact that planes now flew faster and higher, up to 5km (3.1 miles) and more, rapid fire automatic guns of small calibre were still in demand. This caused Krupp to add another weapon to the arsenal consisting to date of the 3.7cm (1.5in) hand-cranked Hotchkiss and selfloading Maxim cannon. This almost unknown gun had been designed to serve as a defence of the *Zeppelin-Luftschiffe*, the dirigibles named after their inventor, Graf von Zeppelin, against British fighters. It was also of the usual small weapon calibre of 3.7cm (1.5in) agreed on in the Petersburg Convention as the lower limit for explosive shells. With an L/14.5 which gave only a very short barrel/tube of 0.5m (1.5ft), showing that the gun was designed for low weight and handyness rather than for hitting power, it still managed to fire its solid shot with a tracer with a muzzle velocity of 350m/s (1,050ft/s) up to 2.2km (1.4 miles), and this with 120 rounds a minute.

Another gun grounded from a plane was the 2cm (0.8in) automatic Becker gun, which later turned up in other plans looking at the possibility of mounting it on a tank. In the end it went to Switzerland after the war, together with its father, and was developed further and returned to Germany as the *Oerlikon-Maschinenkanone*, firing from the nose of the Me 109 through the hollow airscrew shaft in the role of a *Motorkanone*, engine gun, in the Battle of Britain. As a Flak, the Becker gun was no great success, having gotten used to working in the horizontal position in its aircraft gun role, a habit which apparently is very hard to overcome.

All these small calibre guns were only temporary solutions. The *Kogenluft* had put out new demands for the following:

- guns with a muzzle velocity of at least 1,000m/s (3,000ft/s)

*More makeshift German antiaircraft guns were made available by installing even howitzers like this 10.5cm LFH 16 on a 4-ton truck, shown here captured on a French postcard.*

## German Artillery at the Beginning of World War One

*The famous 8.8cm Flak has finally taken on a well-known look with its* Kreuzlafette, *the cruciform mounting for its pivot. But it is still not the model of World War Two, the 8.8cm Flak 18, 36 and 37.*

*(Below) Longer calibres were either mounted on rail cars like this battery of 10.5cm Rheinmetall guns, or mounted stationary.*

- automatic guns with a calibre of 3.7–5cm (1.5–2in)
- shells with a dim tracer, starting at 500m (1,500ft) and lasting up to 4km (2.5 miles).

Not all of these demands had been fulfilled by the time the war ended. Krupp had two different 3.7cm Flak under development, and Rheinmetall a 5cm (2in) Flak.

Shortly before the war ended, a wise decision was taken to reduce the number of different guns of the Flak to the following:

- 7.62mm (3in) for the medium Flak on motor cars
- 8.8cm (3.5in) to replace the 8cm
- 10.5cm (4in) for the heavy Flak (fixed)

All other calibres had to go. The Flak entered the war with 14 Bak. When the war ended it had 2,576. During the war the enemy attacked German territory 1,154 times. Flak guns were deployed as defence (only into a zone of 500km (300 miles) depth): in 1915 150 guns were deployed, in 1916 410 guns, and in 1918 about 900 guns. In 1914–15 these downed fifty-two enemy planes; in 1916 323 planes; in 1917 476 planes; and in 1918 748 planes. The amount of ammunition used for each enemy plane declined from 11,585 rounds in 1914–15; 9,889 rounds in 1916; 7,418 rounds in 1917; and finally to 5,040 rounds in 1918, less than half that of 1914. This shows that the Flak improved both in the field of its weapons and in its

## Antiaircraft Artillery

| Gun model | Calibre (in) | Weight Empl. (lb) | Tube Length (in) | Shell Weight (lb) | Muzzle Velocity (ft/sec) | Max. Range (ft) | Elevation/ Azimuth (degr.) | Remarks |
|---|---|---|---|---|---|---|---|---|
| 2cm Becker MK | 0.8 | 52.5 | n/a | 0.25 | 1,440 | 6,000 vert. | +90/360 | 400 rpm |
| 3.7cm Revolver K | 1.5 | n/a | 47.6 | n/a | 1,620 | 7,500 vert. | +80/360 | 40 rpm (1) |
| 3.7cm Maxim K | 1.5 | n/a | n/a | 1 | 1,620 | 7,500 vert. | +80/360 | 250 rpm (2) |
| 3.7cm Sockel-Flak | 1.5 | n/a | 21.5 | n/a | 1,050 | 6,600 vert. | +80/360 | 120 rpm (3) |
| 8cm Flak | 3.2 | n/a | 144 | n/a | 2,145 | 20,550 vert. | +70/360 | Krupp |
| 8.8cm Kw-Flak | 3.4 | n/a | 158 | n/a | 2,355 | 20,550 vert. | +70/360 | (4) Kr + Rh |
| 10.5cm Kw-Flak | 4.2 | n/a | 189 | n/a | 2,160 | 22,050 vert. | +70/360 | Krupp |
| 10.5cm Flak Kw/E | 4.2 | n/a | 147 | n/a | 1,740 | 17,400 vert. | +70/360 | (5) Rheinm |

As additional antiaircraft guns the following were used: German 7.7cm Fk 96 n/A; 9cm K 73/91; 10cm K 04; captured 7.5cm French FK 97; 7.62cm Russian FK 00 & 02; Italian 7.5cm FK 06 & 11.

*Remarks:* (1) hand-operated Hotchkiss design built under licence by Gruson, firing a cartridge of 37mm calibre with a 94 mm long case; (2) full automatic machine cannon built by DWM under Maxim license, firing a cartridge of 37mm × 101; (3) Krupp design for defence of Zeppelin airships; (4) Kw = *Kraftwagen* = mounted on trucks; (5) E = *Eisenbahn* = mounted on railway cars.

n/a = data not available.

training. Incidentally, the number of enemy planes shot down by German fighters was four times that of the Flak.

The Flak was also asked to help out the infantry against their seemingly invincible enemy on the battlefield in World War One: the tank. Starting in October 1917 at Cambrai, the Flak destroyed no less than fifty-two tanks, but this was not nearly enough.

This role was once again called for in World War Two, when the 8.8 Flak 18/36/37 was employed in the ground role from the beginning, as it was in the Northern Desert. But it could not help the German cities dying in the *Feuersturm*, with about 300,000–400,000 humans being grilled alive in Dresden alone in February 1945, at a time when the war had been decided already.

Today missiles have taken over most of the Flak role against planes. There are, however, some small-calibre automatic guns left, such as the 35mm twin on the antiaircraft tank Gepard.

## FORTRESS ARTILLERY

When the frontier between Germany and France was fortified by both countries from 1872 onwards, they put the guns of the day up in the open-air emplacements on the ramparts, or down in the dark casemates and caponniers. But armour protection for land guns had been thought about since warships had used it in the Crimean War. Thus the first step of the brilliant British Captain Coles, to put the guns inside an armoured turret and this onto a ship, led to the second step of the no less brilliant Belgian Captain Brialmont, to plant one of the British-made Coles turrets, shaped like a cylindrical can, onto the mushroom-shaped redoubt of fort number 3 of the inner ring of Antwerpen, a circle of forts designed by Brialmont himself. That was the day when the artillery received an umbrella against the hail of shrapnel and splinters, and saw the birth of the *Panzerartillery*, which half a century later would start crawling across the battlefield in the form of both tank- and self-propelled guns. In those days it meant that instead of 30–50 guns on the open ramparts, the profit of the armour manufacturers (not Krupp, who came later, but others such as Gruson) would see to it that a fort could only afford two or three under armour cover.

The design was the subject of long discussions among authorities such as Brialmont, the self-appointed world champion, and rivals such as the two ex-soldiers Maximilian Schumann in Germany and Commandant Mougin in France. Even a great test firing near Bukarest in 1885 was unable

*Fortress artillery started at 3.7cm calibre and consisted of the 3.7cm revolving gun, which Gruson manufactured under Hotchkiss licence, first for the navy as defence against torpedo boats, and then in 1874 for the fortifications, to defend the ditches of the forts. Shown here is the 3.7cm RK in the mounting for use on the ramparts, which was the carriage of the old 9cm gun* (left) *and with open rear cover to show the revolving mechanism* (below left).

to decide whose design was superior, since both nations claimed to have won.

There was, however, a real winner at that time: the new shell filled with modern high explosive, which replaced lazy old blackpowder. After this, armour had to be rolled from Krupp's nickel steel; casting it from Gruson's *Hartguss*, hard cast iron, was no longer good enough. This saw the end of both Gruson and the era of masonry forts.

This was followed by concrete and the distribution of the guns in the new dispersed arrangement of the *Festen*, the firmly fortified, which started in 1894 at Metz, where the fortifications were no longer of the *Einheitsfort*-type, the unity forts, which housed both artillery and infantry within one single ditch and wire obstacle. Now there were armoured batteries, usually with four guns in armour cupolas, and for the infantry their own infantry positions, with concrete trenches running around them. All of this was connected by underground posterns to the large multistorey barracks, also buried underground, as were power stations and anything else necessary.

Unlike the first curious mobile foreign designs, such as the French Mougin cupola which kept turning even when firing, the theory being to protect the vulnerable embrasure from enemy projectiles, and were powered by eight unlucky soldiers sweating on long cranks, or the successor by Galopin, which popped up for firing and then sank down again into the protection beneath (a habit preserved in the design of Maginot-line cupolas later), the German *Panzertürme* remained upright and simply slugged it out. But their father Schumann also brought forth some very refined creations. One of them was the *fahrbare Panzerlafette*, the mobile armour carriage, later shortened to *Fahrpanzer*, mobile armour, which was a 5.3cm (2.1in) quick-fire gun mounted in an armoured hemisphere sitting on top of a cylindrical housing for the two gunners. This precursor to a tank was wheeled on narrow gauge rails into its prepared concrete emplacement, or even moved along the roads on a horsecar. It can be debated as to whether this solution, or that of the 15cm (6in) *Kanone in Schirmlafette*, gun in protective mount, which could be moved from one emplacement to another if need be, and of which only twelve were built for Metz, was the one to give birth to the modern self-propelled *Panzerartillerie*. Both depended on other prime movers which were not integrated – only the *Fahrpanzer* could fire its (light) gun from the car transporting it; the powerful 15cm gun would have made the four-wheel trailer it was pulled on by a steam tractor take a somersault.

The standard armour cupolas were shaped like turtle backs, with their artillery inside – 15cm (6in) *Panzerturmhaubitzen*, armour turret howitzers, in the older model of 1893–95 (there had also been two earlier experimental batteries with 21cm (8in) *Turmmörser*, turret mortars, with steel-lined bronze tubes) and 10cm (really 105mm/4in) *Panzerturmkanonen* in the newer model of 1900, both styled on the ideas of Schumann. The tubes moved inside the embrasure on a steel ball screwed onto it when elevating, with the recoil forces on the non-moving gun being taken up by a sort of trunnion on both sides of the breech end, which was closed on firing by a vertical sliding breechblock. Azimuth was taken by lifting the whole 37,500kg (82,500lb) of the turret by means of a tremendous nut and bolt construction turned by hand and levers; the whole turret with the gun rested on a thick shaft ending down in the nut, thus resembling a gigantic mushroom. Then, when it was free from the *avant cuirass* by about 2in and friction contact had ended, the turret was cranked by hand to another direction and the gun was fired. This made the turret swing back and forward, a movement dampened and soon stopped completely by a set of strong Belleville disk springs. The idea was that this would not crack the bond between *avant cuirass* and surrounding concrete, based on troubles experienced with contact between early cement and iron in fortifications. Of these armour cupolas, about 100 were built, mostly in fortifications along the western frontier. The east saw only one battery at Posen (now Poznan).

When Kaiser Wilhelm was told that the cost of one *Panzerbatterie*, armoured battery, amounted to three million Goldmark, he suggested a more

*German Artillery at the Beginning of World War One*

*The 3.7cm RK was soon superseded in 1890 by a single-shot gun of 5.3cm, with a very quick-falling block action invented by Gruson. This gave almost the same rate of fire and contained more canister balls due to its greater calibre. It was called the 5cm, and a close twin in 5.7cm for disappearing cupolas was referred to as the 6cm, to avoid mixing them up.*

*(Below) This same 5cm gun was also installed in the Panzerlafette of Schumann, or as it was later called, the Fahrpanzer. Two men served the gun in its armour housing, which was kept under cover during shelling and only rolled into the open when the enemy started its attack. It was sold worldwide and represented an important step on the way to the automatic self-propelled armoured artillery.*

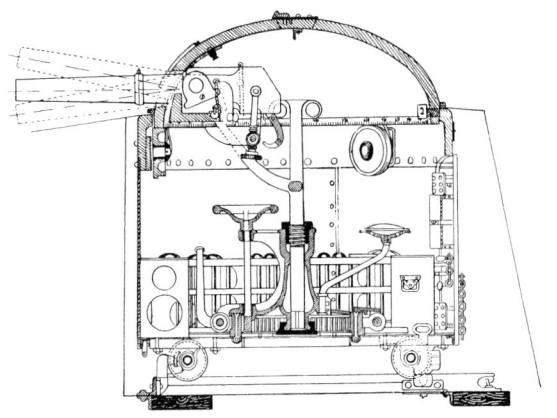

economic version based on his general preference for the navy. This was to be the same 105mm Krupp gun, breech and all, but this time with a system of short recoil, made possible by the tube and the upper part of the carriage recoiling against the brake effect of hydraulics. The whole sat inside a sort of naval gunhouse, which was now also protected at the back,

99

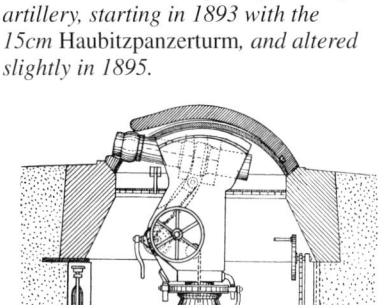

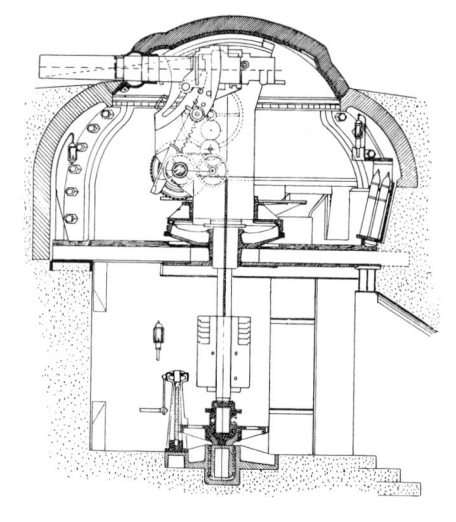

*(Left & below) Also of Schuman design – and Gruson manufacture – were the armoured cupolas for heavy artillery, starting in 1893 with the 15cm* Haubitzpanzerturm, *and altered slightly in 1895.*

*(Left) The 15cm* Haubitzpanzerturm *was followed by the 10cm* Panzerturmkanone *of 1900.*

*(Below) By order of Kaiser Wilhelm II, a cheaper solution was found for armouring the fortress artillery. The 10cm* Kanone *was now installed in* Schirmlafetten *(SL), protected mountings, of the naval type; these SL were now also closed on the rear side.*

and into which the crew of two gunners was able to squeeze. The whole well-rounded gunhouse stood about 1m (3ft) high on a steel pedestal. This arrangement provided a little bit of cover for the loaders, rushing from the gun to the magazine to replenish the 105mm cartridges, since the gunhouse itself could only hold ten of them. It was a nice idea, but even though armour thickness had now dropped a lot from the former 300mm (12in) to 40–80mm, the costs had not fallen to the same degree: a battery of these *Schirmlafetten*, protective mountings, still cost two million Goldmark.

Close defence relied on rapid firing small-calibre guns in the caponniers. After 1872, first old

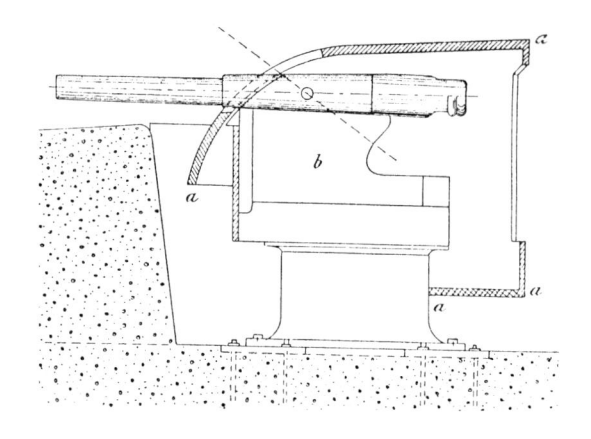

## German Artillery at the Beginning of World War One

bronze smoothbores, former muzzle-loaders converted to breech-loading, had seemed sufficient, even staying smooth (which the troops preferred, as like a shotgun this gave a better pattern with the canister shot – a sort of gigantic grape – fired from them against unlucky attackers). Later they too were rifled and even later superseded by everyday C/61 or C/64 guns, which received a special mounting, the *Kasemattenlafette* C/73, casemate carriage of 1873. This was styled after the mountings of the big coastal guns of those days, with the gun recoiling in the upper part of its mounting braked by a hydraulic cylinder and impacting at the end on rubber blocks.

Later doctrine moved away from the big shotgun toward a small rifle that fired rapidly. This was in 1884, when Gruson was able to sell the 3.7cm (1.5in) hand-operated revolving five-barrel gun that he was manufacturing under Hotchkiss licence, to the *Ingenieurkomitee*, the Engineers' Committee, responsible for the design of German fortifications. These weapons had, as we have already learned, been ordered by the *Kaiserliche Marine*, which like all other navies was nervous about the *schwarze Gesellen*, the black fellows, the name given to the crew of the new torpedo boats, normally covered in the soot of their funnels. They were too fast to be hit by the heavy but slow main artillery of 9–10in. So the small-calibre quick-fire guns were sent to do this job.

By now, however, the torpedo boats had grown and 3.7cm shells were no longer absolutely fatal. They were still used for keeping the ditches clean, even though the *Revolverkanonen* fired canister against troops, the ditches of the forts being just the maximum length (300m/900ft) for using this. In spite of all sort of tests and favourable reports received before mounting these revolving guns in the forts in 1884, it took only four years for the complaints of the unsatisfied customers, the gunners working them inside the forts, to reach the ears of the APK. Now the Gruson-Hotchkiss was too noisy and spewed its blackpowder fumes into the caponniers. There was some truth in this: the barrels were shorter than the distance to the wall outside; as a matter of fact they ended halfway inside the wall, but they had not been any longer during the extended tests which had been satisfying. Still, the troops got a new gun, which Gruson just happened to have finished inventing at that time: the *Deutsche* 5cm (2in) *Schnellfeuerkanone*. This designation was to distinguish it from the 6cm (2.4in) *Kanone*, which was the same gun but with a calibre of 57mm (2.3in). The latter was only found in the vanishing close defence turrets of the two forts at Mutzig near Strasbourg, which formed the nucleus of the later *Feste Kaiser Wilhelm II*, based on the new vertical sliding breechblock, invented by Gruson and adapted by the APK. The name *Schnellfeuer*, quick-fire, was used because the gun fired fixed cartridges, which contained everything in the cartridge case: projectile, propellant and primer. These guns were in use until the first machine guns arrived. These were of the ubiquitous Maxim design and their last model was introduced into the German army in 1908 and therefore called MG 08.

Hundreds of both the 3.7cm Gruson revolving gun and the later 5cm quick-fire gun, which had also served in the *Fahrpanzer*, were stored lovingly in the depots and waited just like Kaiser Barabarossa (Emperor Friedrich I, 1152–1190, called Barba Rossa by his Italian subjects because of his red beard, who was said to be sleeping in the Kyffhaeuser mountains) until the time came to retrieve them. This was in 1914 and they then both served as antiaircraft guns, with the 3.7cm revolving guns later fathering an offspring of antitank, trench and infantry guns. The other heavy guns of 12cm and 15cm (4.8in and 6in) calibre, which formerly stood on the ramparts, had been retired to the relative safety of the intermediate batteries, leaning themselves against the flanks of the fort and using this now only as quarters for the gunners and as an ammunition depot. The only guns remaining up there as targets were about 100 former 15cm (6in) long ring guns in coastal mountings, which the navy had handed over to the land fortifications.

The German fortress artillery fulfilled its designated role: it kept the enemy away. Only one incident of their artillery having to fire in World War

101

*German Artillery at the Beginning of World War One*

*A whole battery with four of these 10cm KiSL has survived in the* Feste *Kaiser Wilhelm II at Molsheim/Mutzig near Strasbourg, on French terrain since 1918. Since this* Feste *also features two forts armed with 15cm HPT and a whole battery of 10cm PTK, together with a very impressive museum, it is well worth a visit. The modern type of German fortifications, the so-called* Festen, *built from 1893 onwards along the borders of East and West Germany, also featured special casemates with artillery to close the intervals between the different forts of a fortress ring around a city like Metz or Diedenhofen (Thionville). There, 7.7cm versions of the FK 96 n/A, but with a shorter recoil, were paired behind 10cm of armour plate, very often with a searchlight on top.*

One is reported. This was in August 1914 when the French, descending from the Vosges mountains south of Strasbourg, clashed with the German defenders. This gave the artillery of the *Feste Kaiser Wilhelm II* at Mutzig the chance to free their tubes of rust, but that was it until 1918.

In World War Two, when the fortifications built by the Germans were again manned by the *Wehrmacht*, the test was harder. The thrust of the US troops was stopped for a quarter of a year by the ring of *Festen* around Metz, which had been built around 1900. Artillery made little impression on the concrete works; neither did bombing. Attacks by tanks were stopped by defenders popping up out of exits from the underground infantry restrooms, which were some 20–30m (60–90ft) behind the concrete trenches, and knocking off the Shermans with their *Panzerfaust*, the hand-held recoilless antitank weapon. And when the GIs managed to get into one of these tunnels and started an attack along the subterranean posterns and tunnels, it was soon stopped. As General Nicolas of the French *genie*, the engineers, who had served as an adviser on the staff of Patton, told the author: 'You came along the postern and there was the door of iron bars closing it. And 10m (30ft) behind this door was another one, this time made of armour, with a loophole. And an MG 42, the standard 8mm machine gun of the *Wehrmacht*, sticking out of this embrasure waited for anyone to come near the door with the iron bars. So it was the end.'

Despite building literally thousands of bunkers – the Westwall alone numbered over 15,000 – there was no genuine German fortress artillery with armour cupolas in World War Two. The only guns installed in such were some 150 5cm *Maschinengranatwerfer* M 19, automatic launchers firing at a rate of 120 rounds a minute, and half a dozen 10.5cm (4in) *leichte Turmhaubitze*, light turret howitzers installed in the *Atlantikwall*, the atlantic rampart, near Boulogne. All other plans for giant armour cupolas weighing in at *Baustärke* A, strength of construction A, which meant ceiling and walls made of 3.5m (10.5ft) of concrete with armour cupolas 600mm (2ft) thick, around 600,000kg (1.3 million lb) in total, were prepared, especially for the *Ostwall*, the fortifications in East Germany (Pologne today), but were hardly begun before the war started and were not finished. Today, of the over 20,000 *Westwallbunker* (15,000 until 1940 and later an additional 5,000 in 1944–45), not to mention those of the *Atlantikwall*, *Ostwall* and *Südwall*, the southern rampart around the Mediterranean Sea, only around 500 have survived in Germany, most of them in the southern area called the *Saargebiet*, which the French hoped to get after 1945. Surprisingly, the population opted to stay with Germany when asked officially for their decision. Today the open frontiers show that it does not take an army to go from Germany to France or the other way; common sense and *fraternité* are enough.

**Fortress Artillery**

| Gun model | Calibre (in) | Weight Empl. (lb) | Tube Length (in) | Shell Weight (lb) | Muzzle Velocity (ft/sec) | Max. Range (ft) | Elevation/ Azimuth (degr.) | Remarks |
| --- | --- | --- | --- | --- | --- | --- | --- | --- |
| 3.7cm Revolver-K | 1.5 | 1,256 | 31.8 | 1 | 1,200 | 6,000 | n/a | (1) Gruson |
| 5cm Kasematt-K | 2.1 | 1,245 | n/a | 3.67 | n/a | 9,000 | n/a | (2) Gruson |
| 5cm K i.Pz.Laff. | 2.1 | 5,283 | n/a | 3.67 | n/a | 9,000 | n/a | (3) Gruson |
| 6cm K. i. Senkturm | 2.3 | n/a | n/a | n/a | n/a | n/a | n/a | (4) Gruson |
| 7.7cm Kas K | 3.1 | n/a | 84.1 | 14.4 | 1,395 | n/a | n/a | (5) Krupp |
| 10cm Turm-K L/35 | 4.2 | 82 tons | 14 | 41.2 | 1,680 | 33,200 | +35/360 | (6) Krupp |
| 10cm KiSL | 4.2 | 3,300 | 141 | 41.2 | 1,680 | 33,200 | n/a | (7) Krupp |
| 15cm K iSL | 5.9 | 27,192 | 236 | 115 | 2,250 | 46,800 | +35/125 | (8) Krupp |
| 15cm RingK L/23 | 5.9 | 6,694 | 136 | 61 | 1,455 | n/a | n/a/360 | (9) Krupp |
| 15cm lgRingK L/30 | 5.9 | 13,233 | 180 | 61 | n/a | 30,000 | n/a | (10) Krupp |
| 15cm Turm-H L/11 | 5.9 | 2,490 | 49.5 | n/a | n/a | 2,160 | n/a/360 | (11) Krupp |
| 21cm Turm-H. L/12 | 8.4 | 4,783 | n/a | 200 | n/a | 1,275 | +40/360 | (12) Krupp |

*Remarks:* (1) Same Gruson-Hotchkiss hand-operated revolving gun 37 × 94 as in antiaircraft guns, but in casemate mounts or rampart carriages; (2) Gruson design of 53mm calibre with a hand-operated vertical breechblock in casemate carriage; (3) same 53mm gun in a mobile armour housing of Schumann design; (4) same gun as above, only now in 57mm calibre, mounted in a disappearing turret of Gruson design; (5) FK 96 n/A with shorter recoil for armoured casemates; (6) 105mm cannon without recoil system, mounted in an armour turret of Schumann design, firing cartridges, weight of cupola alone; (7) 105mm cannon with short recoiling upper part of carriage, mounted in a naval-type, but fully enclosed armour housing, was removed from the fortresses and placed elsewhere in batteries, weight of gun alone; (8) 150mm gun in armour box open in rear, mobile after dismantling from position, all transported to front; (9) ring gun mounted in minimum embrasure carriage, in pairs inside Gruson hardcast armour cupola and also on field carriages without recoil system outside; (10) old ex-coastal gun in frontal pivot coast carriage mounted on the ramparts; (11) 150mm howitzer without recoil system mounted in armour turret of Schumann design, weight of howitzer alone; (12) 210mm mortar with steel-lined bronze tube, mounted without recoil in armour turret of Schumann design, weight of weapon alone.

n/a = data not available.

## COASTAL ARTILLERY

After the Napoleonic wars, Germany did not exist as one united nation, as the Holy Roman Empire of German Nation had been disbanded in 1806. Prussia had come out of the wars of liberation, as they were called, as the leading nation in the *Deutsche Bund*, the German Union, but this included the multinational Austria. This was the reason why Chancellor Bismarck later sought the *kleindeutsche Lösung*, the small German solution, of a purely German nation without Austria and its many Balkan acquisitions promising trouble in the future. The war with Denmark in 1864 for Schleswig and Holstein in the north still saw the Austrians at Prussia's side (two years later they would be on opposite sides) and this was fortunate, as only Austria had a fleet of warships. The Prussians did not, and when the Danish warship *Rolf Krake*, built in Britain on the lines of the US *Monitor*, threatened to sail into the Flensburg harbour, Prussian 24-pounders (15cm/6in) C/61 of the siege train had to drive her off, since there was simply no coastal artillery in the whole of Germany.

The neighbouring port of Kiel was defended by the makeshift means thought up by a brilliant young lieutenant of artillery: a chain of barrels filled with blackpowder floating across the entrance, each connected by waterproof wires to a galvanic battery, forming an observed minefield. The electricity would gain the young man both worldwide fame and an industrial empire: Werner von Siemens. After the 1866 war Prussia found herself the owner of various territories of German nations unwise enough to have backed the wrong horse – Austria – and now having to pay for the lost wager. These included long strips of coast in northern Germany, terrain which would have to be defended, as there was also the Eldorado of rich ports such as Hamburg. So Moltke made a study of the situation and Prussia started building coastal forts, with Krupp furnishing his range of big-calibre naval guns, only this time on coastal mounts.

The first forts were built to close the approaches to the ports lying on the rivers Weser and Elbe, which could be reached from the sea. The Weserforts were open-air museums of the development of coastal artillery, ranging due to the time of their construction from open-air 21cm (8in) batteries (Brinkamahof I on the right bank of 1867) and enclosed batteries of hard cast iron (Langlütjen I on the left bank of 1869). There were also 21cm Ringkanonen L/22 and Langlütjen II (1872) up the river on an artificial island near the left bank, armed with five Gruson armour turrets with a single 28cm (11in) ring gun L/22 each, and finally Brinkamahof II (1875) upriver on the right bank with three Gruson turrets made of hard cast iron and armed with the popular German calibre of 28cm, as well as *Ringkanonen* with tubes of the prestressed construction invented by Armstrong, but this time in twin mounts. All of the L/22 and later L/30 guns used blackpowder as a propellant. The coastal fort of Kugelbake near Cuxhaven, closing the River Elbe towards Hamburg, did not receive as much money and therefore had to leave its 28cm ring guns L/22 in open emplacements, even in 1879 when it was finally finished. But then came 1890 and the British swapped the island of Heligoland, which they had acquired by taking it from the Danes in 1807, for the tropical island of Sansibar, which Germany had acquired in classic empire-building tradition by taking it away from the natives. Helgoland (in German the extra 'i' is missing) was ideally placed for the new German navy, lying only about 60km (35 miles) ahead of the German coast in a watchdog position. All the world agreed that this exchange was for the British what buying Alaska from the Russians had been for the USA: a fantastic stroke of luck. Of course, today one may see things differently, even if only remembering that the Germans had all their colonies taken away by the victors after World War One, having been declared unfit to lord it over poor natives, whereupon these same colonies were split up among the superior victors.

Thus Helgoland was fortified at once, this time by the navy. One of the first things navy-lover Wilhelm II had done when ascending the throne in 1888 had been to make the army hand the job of coastal defence over to the navy. This was doubtless a well-founded decision, setting a thief to catch a thief, as

*German Artillery at the Beginning of World War One*

*Prussia had been a continental power until it acquired the northern terrains after the 1866 war, which had found Hannover on the losing side. In the 1870s the River Weser was protected downriver from Bremen by forts with first open and then armoured batteries. These featured Krupp ring guns from 21cm up to 28cm.*

*(Below) Krupp also supplied guns in medium calibre on disappearing mounts.*

the army artillery was not used to its targets moving when being shelled. The navy, on the other hand, was. Already in 1903 the first batteries on the high part of the island, called *Oberland*, high country, in recognition of the fact that it was 50m (150ft) above sea-level, were finished. They consisted of eight of the old high-angle mortar type (28cm/11in *Küstenmörser*, coastal mortars, now called howitzers L/12) and four flat trajectory low-angle type (21cm/8in guns L/35 behind well-rounded armour shields, formed in the *Schirmlafetten*-style then popular). To these were added in 1912 eight of the really powerful 30.5cm (12in) L/50 coastal guns, mounted in pairs in four twin turrets. Like all the other guns, these were of course of Krupp design, with a front armour of 40cm (16in) that no enemy shell was expected to pierce. A long tunnel connected all the heavy gun positions, with dozens of medium- and small-calibre quick-fire guns added to the searchlights and range finders along the beaches, and from 1914 onwards a countless number of antiaircraft guns were distributed over the island. All of these were going to prove Clausewitz's point: that creating an impenetrable position on the front entrance forces the other fellow to look at the lock of the back door.

Contrary to this, however, shots were fired in anger from German coastal fortifications at the beginning of the war. These were heard at Tsingtao in China, where Germany had leased terrain, not for her part in suppressing the Boxer rebellion in 1900, but through a simple peacetime contract with the Chinese Empress in 1898. The port of Tsingtau had then been fortified like Helgoland. But it was attacked and then taken by the Japanese at the outbreak of the war. They were certainly no Slavic bloodbrothers of the Serbs, like the Russians, but saw only a good opportunity to kick a man when he was down and then rob him, a well-known practice in international diplomacy in the years before the UN. The fortifications of Tsingtau were no match for modern warship artillery and Germany did not plan to send relief to somewhere so far away. So it was one for the warships.

The coastal guns made good at another place. The German thrust through Belgium had ended in October 1914 on the Flemish coast of the Channel, with Britain on the other side of the water. Britain was known worldwide for its superior fleet, which

105

*German Artillery at the Beginning of World War One*

(Above) *The decade of the theory of high-angle fire against the thin decks of naval targets did not spare Germany, as shown by this 28cm* Küstenhaubitze.

*Medium calibres featured 10cm guns in* Schirmlafette, *the naval-type open in back, and also 17cm guns preserved from old battleships.*

*When World War One reached the Flandrish coast, it brought with it a lot of guns. Among these were the 10cm Krupp* Küstenkanone *L/50 in* Räderlafette, *coastal gun in wheeled carriage, with its incredible 19.5km range. It is shown here on a circular bedding* (top) *and with pedrails* (bottom).

ruled the waves, and even though *Kapitänleutnant* Weddigen had put some fear into them with his U 9, it did not stay in the ports for long. Besides landing the soldiers of the British Expedition Force (BEF) in France, the fleet also tried to push the Germans out of Belgium and to put a cork in the bottleneck at Seebrügge to stop the submarines from exiting there. This resulted in two remarkable things: the impressive design of an 18in gun as top gun on British monitors for shelling German positions, especially at Ostende where the German submarines were also stationed while being served up the Zwine River at Brügge, and later in the first of the submarine pens destined to gigantic growth in World War Two; and the build-up of German coastal defence in Flanders to protect this fleet. All sorts of guns were sent on the march for this, both from the army and the navy. Even venerable antiquities such as the 28cm L/22 blackpowder ring guns of Fort Kugelbake were not left in peace at home, but had to limp along to the coast in Flanders. There all guns were emplaced in various different positions, firstly firing from behind a parapet in the classic way of the last five centuries, then lowered into the circular concrete rings of the *Kesselbettungen*, kettle beddings, where they stayed in the open air or were later protected by the armour of gun housings.

For heavy batteries such as Tirpitz, south of Ostende, the swampy ground had to be rendered

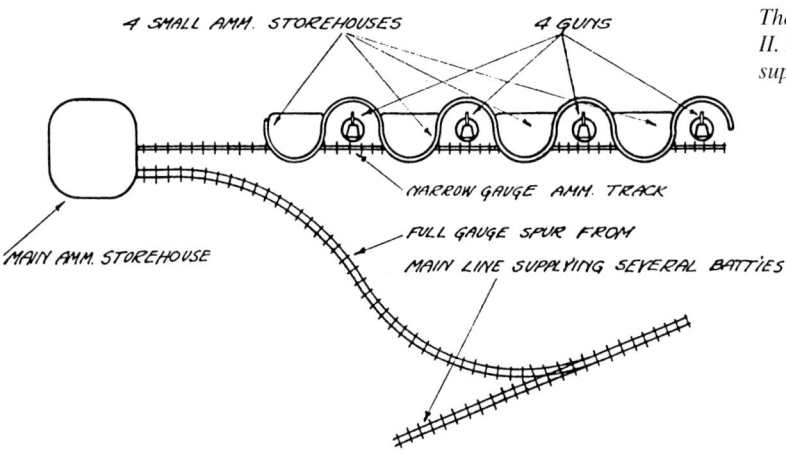

*The 30.5cm battery Kaiser Wilhelm II. Shown here is the ammunition supply.*

stable to bear the concrete emplacement by driving thousands of piles into it. Also, considerable attention had to be paid to its drainage and that of the Oldenburg battery. The ammunition was stored in concrete bunkers, the shells on the right traditionally – as they are even today in all artilleries of the world – separated from the propellant, the powder. There were projectiles of three different weights and the powder charge consisted of the traditional *Vorkartusche*, front charge, in bags, followed by the *Hauptkartusche*, primary cartridge, the main charge, also referred to as *Hülsenkartusche*, encased propelling charge, consisting of more bagged powder ending in a brass case with the primer. The shells, both explosive and semiarmour-piercing, all had a false nose/ogive acting as an aerodynamic windshield, which increased range. The ammunition was pushed on handcarts from the store to the gun. In the case of the old high-angle mortars also installed in Belgium in deep concrete pits of 6m (18ft) diameter behind a shielding concrete parapet 5m (15ft) high, as in the 28cm *Batterie Groden*, the charge was low enough to be carried by two men on a tray.

The emphasis on Seebrügge and Ostende resulted in a chain of all existing calibres of naval and antiaircraft guns, with the heavies sitting in their concrete shelters behind the protecting dunes, firing indirect fire over them at enemy ships according to the range and azimuth received from the observation posts with their long-distance range finders erected on the dunes. The batteries were shielded from direct contact with the enemy by a line of dense field fortification starting at the beach with barbed wire obstacles and continuing with trenches, machine gun positions, searchlights and pill boxes, shades of the *Atlantikwall* to return there twenty-five years later. The guns were manned by two regiments of *Matrosenartillerie*, sailor/naval artillery, no. I in the east sector around Seebrügge and II in the west sector near Ostende. They worked all sorts of batteries, against enemies both on the water and in the air, twenty-three naval batteries being counted in 1918 alone. The calibres of the ten heavy coastal batteries ranged from 28cm (11in) over 30.5cm (12in) up to 38cm (15in) with a range between 25–47km (15.6–29 miles); the twenty-three medium and light batteries sported guns ranging from 8.8cm (3.5in) over 15cm (6in) up to 21cm (8in), with a range between 10–20km (6–12 miles). These had to ward off British attempts to block the exit of the German submarines by sea or air action against the two harbours and their canal locks. Even commando raids were denied success. Before they started firing, a long line of big smokepots were lit along the beach to hide the smoke coming from the tubes and frustrate the counterbattery fire of the British coastal monitors. A network of railway lines had been built to supply the heavy batteries with ammunition, which was also heavy, with the 38cm (15in) armour-piercing shell in the range of 900kg (1,900lb). This was especially true in the case of batteries such as Deutschland, Pommern, Preussen, Hannover or Tirpitz (all except the last one named after geographical terms). Pommern had its guns inside armour gunhouses.

Of course, like everywhere else, these batteries and their guns could not be taken home at the end of the war and so were blown up by their crews in 1918 after having fired until the last minute. This was done by ramming one shell without driving bands high up the tube, then loading another one and firing this with a really long line. The second shell impacted upon the first and detonated both, blowing off the breech piece of the gun. (*Vorrohrsicherheit*, safety of fuses, when travelling in the tube and out of it for about 100m, was not applicable in this case, where the impact of a high-speed shell was enough to set off its explosive.) The only guns that got away were the railway guns. The pillboxes and emplacements were also removed later; it takes about 2,000 years for a people to accept and admire the signs that they had once been conquered and occupied, like looking at Roman ruins today.

There is one exception to this rule of the people making a clean sheet of their history. This lucky scenario occurred near Ostende at the *Domein Ravesijde*. It too was to be swept clean after 1918, but the owner said no. And since he was Prince Charles of Belgium, the apparent heir to the throne (who indeed did reign in Belgium from 1944–1950), he had his way and all the installations there were preserved. In 1940, a German voice cried, '*Alles auf die alten Plätze!*', Everyone into the old positions, and a new battery was built there, later incorporated into the growing *Atlantikwall*. This too was preserved after 1945 (again due to Prince Charles of Belgium) and thus Raversijde has remained a unique sight in the world, an open-air museum of a modern fortification, with shelters, guns, and so on.

Other German coastal fortifications met quite different fates, depending on where they had been built. The *Sicherungsstellung Nord*, securing position north, of World War One in Denmark, built to ward off a feared British invasion, was blown up by the Danes after 1918. So were all World War Two bunkers on the German coast. Belgium and Holland hid them mostly under heaps of sand, creating artificial dunes, with Raversijde again being an exception. The pragmatic French built wonderfully sited houses on them, receiving a remarkably cool wine-

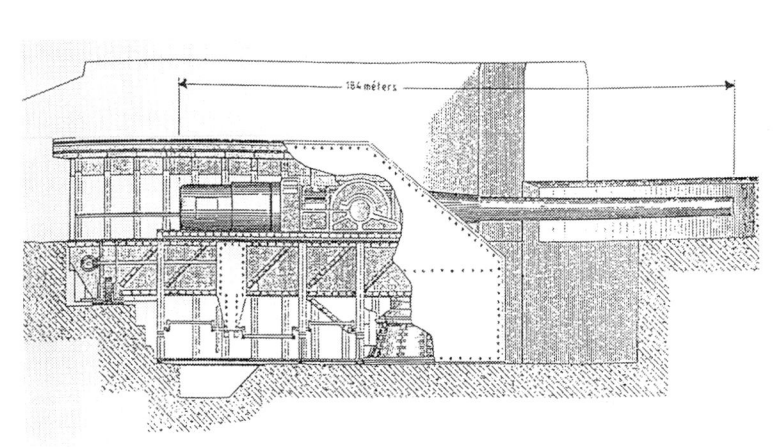

*The 38cm (15in) heavies like the Deutschland and Pommern batteries were shielded in armour.*

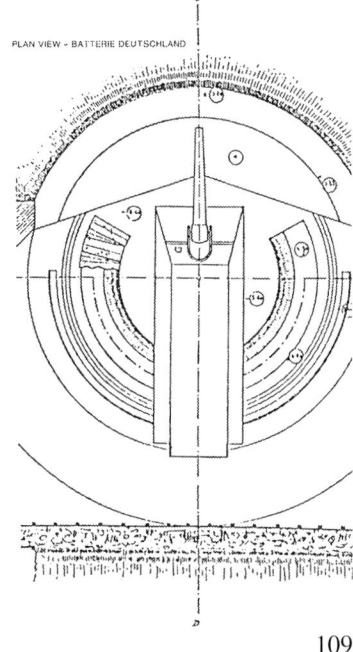

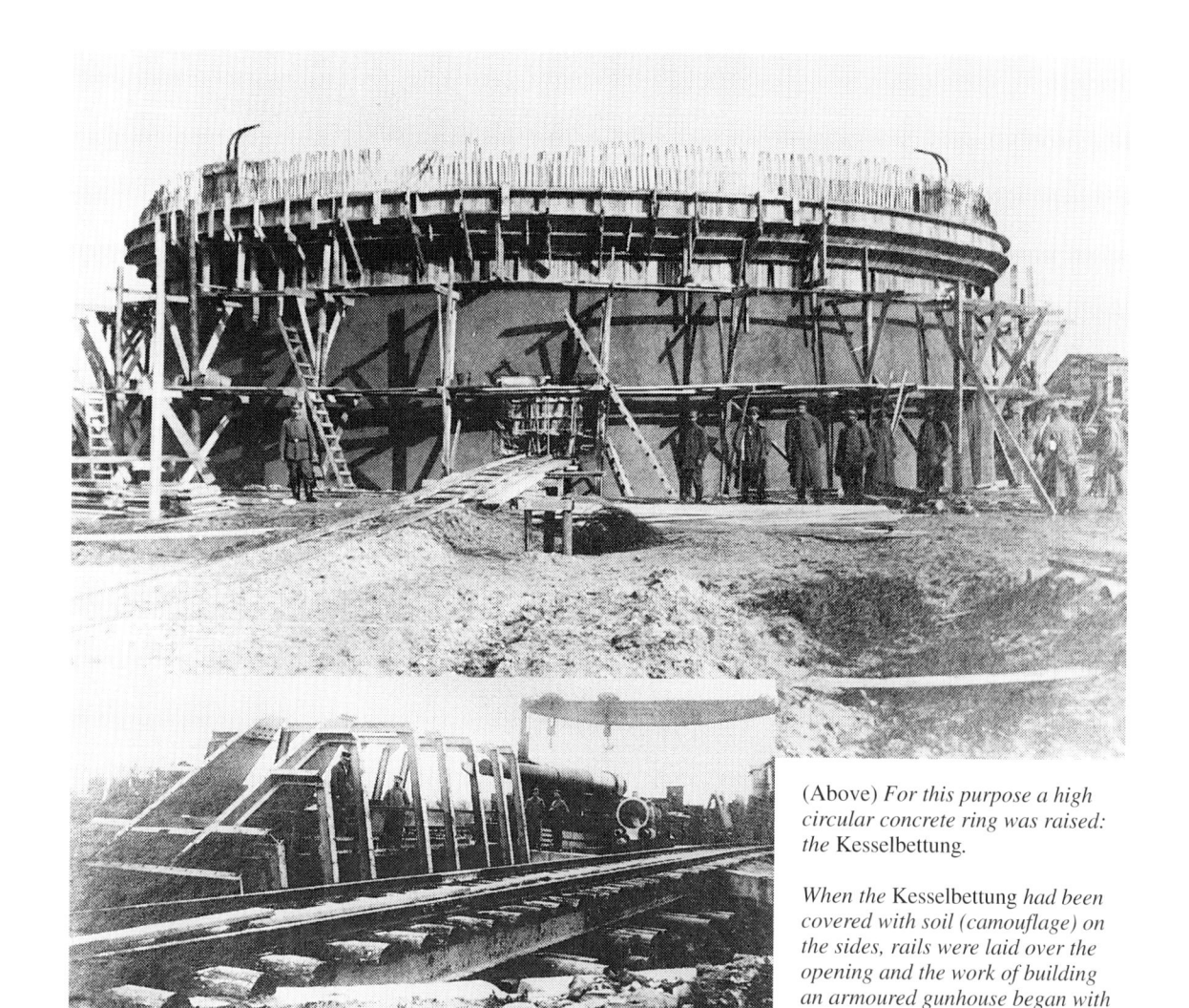

(Above) *For this purpose a high circular concrete ring was raised: the* Kesselbettung.

*When the* Kesselbettung *had been covered with soil (camouflage) on the sides, rails were laid over the opening and the work of building an armoured gunhouse began with raising the strengthened I-beams on the turntable.*

*To these ribs the armour plates were fastened.*

*The mighty gun was pushed through the embrasure left open in front.*

(Above & right) *Now the gun was ready to belch its 1-ton projectile amid flames and smoke.*

(Above) *These shells are impressive enough, but they belong to smaller calibres of 30.5cm or even 28cm (two driving bands only).*

*A 38cm shell with three driving bands, and looking even larger as a small soldier has been found for comparison.*

cellar as a bonus. And the British? Look at the unique Dover turret buried in the Admiralty Pier.

Modern coastal defence no longer relies on sheltered coastal guns. Satellites watch every shovel of concrete used in their building and report their location to the Intercontinental Ballistic Missiles (IBM), which will hit them with a CPE, a dispersion of feet only, not that this matters much with nuclear warheads. Astonishingly, both Sweden and Norway have invested a lot of money and steel in their granite coastlines, but this has not stopped nosy visitors from crawling around at the Swedish naval base Karlskrona and elsewhere in tracked minisubmarines.

*Let us not forget the lesser calibres, like this 17cm* Kanone in Schirmlafette. *Note the telephone operator looking out of his concrete shelter, which also protects the ammunition. The gun is served by a crew of sailors.*

(Below) *Guns of a mere 28cm calibre look impressive, especially when inside an armoured gunhouse.*

*This bunker is built on the same principle as the* Kesselbettung. *First the bunker is erected and covered with soil to hide it from the eyes of the watchful enemy pilot. This work is normally done under cover of camouflage netting. Judging from the wide entrances built for trucks to pass inside, this building will house ammunition, not men.*

*German Artillery at the Beginning of World War One*

*A big gun in an open* Kesselbettung. *The crew necessary to serve the gun (80–120) depended on whether elevation was done by hand-cranking or assisted by electric motors.*

(Below) *There were also railway batteries on the coast in Flanders.*

(Bottom) *Two of the 30.5cm coastal guns in armour turrets firing on Heligoland. They only fired once in anger, on 24 November 1914.*

*A battery of naval artillery, 17cm SK in* Schirmlafette, *watching the sea from the dunes near the Dutch border. In the foreground is the fire control with an observer behind the* Scherenfernrohr *and a 4m rangefinder to give range and azimuth to the guns.*

## Coastal Artillery

| Gun model | Calibre (in) | Weight Empl. (tons) | Tube Length (in) | Shell Weight (lb) | Muzzle Velocity (ft/sec) | Max. Range (ft) | Elevation/ Azimuth (degr.) | Remarks |
|---|---|---|---|---|---|---|---|---|
| 3.7cm Maxim | 1.5 | n/a | n/a | 1 | 1,680 | 18,000 | +80/360 | (1) DWM |
| 5.2cm SK L/52 | 2 | n/a | 108 | n/a | n/a | n/a | n/a | (2) Krupp |
| 8.8cm SK L/35 | 3.5 | | | | | | | (3) Krupp |
| 10cm SK L/40 | 4.2 | n/a | 168 | | | | | (3) Krupp |
| 15cm SK L/45 | 5.9 | n/a | 270 | 92.4 | 2,520 | 68,100 | +45/360 | (3) Krupp |
| 17cm SK L/40 | 6.8 | 49.4 | 272 | 134 | 2,445 | 72,000 | +45/360 | (3) Krupp |
| 21cm SK L/45 | 8.4 | 108 | 378 | 275 | 2,400 | 78,000 | +45/360 | (3) Krupp |
| 24cm SK L/35 | 9.2 | 113 | 336 | 332 | 2,010 | 50,100 | n/a | (4) Krupp |
| 28cm SK L/40 | 11.2 | 129 | 448 | 528 | 2,460 | 88,500 | +45/360 | (4) Krupp |
| 28cm KüstMörser | 11.2 | 73.3 | 134 | 735 | 1,038 | 31,200 | +65/360 | (3) (4) Krupp |
| 30.5cm SK 12 L/50 | 12 | 1,577 | 610 | 850.5 | 2,610 | 85,500 | +30/360 | (5) Krupp |
| 38cm KüstH L/16.5 | 15 | n/a | 250 | 3,120 | 2,100 | 51,000 | n/a | (6) Krupp |
| 38cm SK L/45 | 15 | 240 | 684 | 1,650 | 3,120 | 142,500 | +55/360 | (7) Krupp |

*Remarks:* (1) *see* antiaircraft artillery; (2) laid by moving with shoulder; (3) naval-type with armour shield open in back; (4) mounted on rails; (5) old weapon, mounted in pits: (5) modern coastal gun mounted in twins in armour turret with 16in front ; (6) design started in 1912 for upgunning mortar batteries on Helgoland. One mortar made in 1913, remained at factory until in 1918 conversion for firing from army *Schiessgerüst* started, but was not finished until the end of the war; (7) *Batterie Deutschland*, weight of gun alone without armour.

n/a = data not available.

# 3 Developments of 1916 and Beyond

## LESSONS OF THE WAR IN 1914 AND 1915

Germany's artillery had not been prepared for the new form of war which started in 1914. War had been a grim and unforgiving teacher, his lessons bloody and fatal. For the German artillery they were that guns had to fire further and have more effect on the target. Now range was indeed something the field guns had been lacking from the beginning. But this was not because of a deficiency in the ballistic qualities of the guns. The old tube of the FK 96 of 1896 had had to be used again when modifying this into the newer model FK 96 n/A by simply putting it onto a lathe. But this old tube had been good enough for a range of 7,000m (21,000ft) as far as its ballistics were concerned. This was only 700m (2,100ft) less than the range of the much-admired French field gun, the 75mm M 97, the *soissante quinze*. But neither gun used its ballistics to the full extent. The FK 96 n/A fired at maximum elevation to only 5,325m (15,975ft), when the tail was not dug into the ground. The same was true of the French gun. This could have fired up to 8,000m (24,000ft) according to its tube, but the carriage permitted only 6,800m (20,400ft), the instruments for laying the gun only 6,500m (19,500ft) and the firing tables ended at 5,500m (16,500ft) for the shell.

When the FK 96 n/A was later criticized for its inferior range, it was forgotten that at the time of its design, around 1890, there had been no telephones for contact and exchange of information between forward observers and the battery. So the observation post had had to stay next to the guns. To register the impacts and evaluate the terminal effects from there – hit or miss – was only possible from a distance decidedly shorter than the maximum range, even of this time. Thus the French artillery still held that firing at ranges over 4km (2.5 miles) was impossible, even in 1910.

The FK 96 n/A was superior to the French 75mm M 97 in other ways. The higher muzzle velocity of the M 97 had resulted in a flatter trajectory, which kept it further away from the protective cover over which it had to shoot in indirect fire. The FK 96 n/A was also easier to handle in heavy ground due to its lower weight. The M 97 had to be put on *abbatage,* brake shoes, to gain a solid position. This slowed greater changes of azimuth a good deal. The FK 96 n/A also had a larger armour shield, giving better protection to its crew when accompanying the infantry.

Using single FK far ahead in the first line greatly impressed the French in 1914. After the war the French General of Artillery Herr called the FK 96 n/A an 'excellent gun', up to the standards of the time at which it was introduced into service. The French, however, had to live with certain deficiencies of their M 97. One of them was the flat trajectory leaving too much dead ground underneath. It was hoped to compensate this by an addition named the Malandrin plate, after its inventor. The French had hoped that this would be sufficient to compensate for the light 120mm (4.8in) field howitzer that Schneider-Creusot had designed before the war and offered to the army. The Malandrin plate consisted of a circular plate bolted to the nose of the shell under the fuse. First it was small and rigid, later larger and folded up for loading, being rotated by the centrifugal force into a horizontal position. This enlarged air resistance therefore slowed down the shell and resulted in a more curved trajectory. In the end the French dropped this in favour of reducing the propellant charge in 1917. Still the M 97 was insufficient in this field, as were the (missing) French field howitzers, according to the French General Gascoine, who after the war accused the French field artillery of:

- not having a great enough effect against German field positions
- straining the French industrial capacity by the unwarranted demand for powder, guns and shells
- frequently firing at their own positions because of the flat trajectory of field guns
- lack of cooperation between infantry and artillery.

German soldiers on the other hand were impressed by the sharp bang of exploding French 75mm shells. The impact of the German 77mm field gun resulted only in a sort of muffled plop. The reason for this was that the French, relying on the blast of the explosion for effect, had packed a lot more explosive into the shell (825g/1.7lb) than the Germans, who had put their trust in the more numerous fragments caused by the thicker wall of the shell, which left space inside for only 190g (0.4lb). During the war, both sides demanded and got what the other had: France, in the *obus D*, projectile, reducing the filling to 285–650g, and the Germans enlarging the filling to 900g (2lb).

Germany stayed in the lead as far as high-angle fire was concerned, even after the French army reactivated their old 95mm (3.8in) deBange guns, which in spite of their shells weighing 12.5kg (26lb) were no match for the German 10.5cm lFH, especially when considering that their carriages only took an elevation of up to 24 degrees.

## NEW GUNS OF 1916

The two main demands of German artillery resulting from the experiences gained during the war have been discussed: greater range and more effect.

A solution to the first problem was the responsibility of the gun designers; for the second, more heavy artillery was needed, as was more effective ammunition.

The new field guns introduced during the war from 1916 onwards were not the first step in this direction. The range had already been increased by digging a hole under the trail of the gun and lowering this into the ground, a solution used by both sides, though not a perfect remedy, as quick change of azimuth by a greater amount was now impossible.

Greater range also called for other means of observation and target acquisition. Thus the balloon repeated what it had done in the US civil war half a century earlier: carry the observers up to a height where their view was unhindered by the raises in the terrain, but also where they could be detected by the enemy from afar, with enemy fighter planes sooner or later appearing in the sky. These were no longer simply nuisances, as by now the machine gun had been married to the plane, with all sorts of devices to make it fire from the fuselage without hitting the airscrew and downing the plane. Special incendiary balls containing phosphorus behind the lead tip were guaranteed to inflame the hydrogen filling, giving the observer in the basket occasion to be grateful to the thrifty army for providing a parachute for accidents such as this, something that was not yet the case for the pilots of all other nations.

The first step when increasing the range of a gun is to put in a longer tube. But this was not as easy as it sounds. The short tubes of early field guns could be mounted by their trunnions in the carriage in equilibrium, without needing any equilibrator. Elevation could be taken with the help of a simple screw rod working in both directions. Now a heavier and more complicated tooth quadrant had to be used. In Germany, the desire of most higher officers of the artillery was to do away with digging holes for the gun trail. For a quick measure Rheinmetall proposed mounting the cannon of the FK on the carriage of the howitzer lFH 98/09. This resulted in an elevation of up to 40 degrees and the full range of 7.8km (4.9 miles) that the present tube could produce.

## The New Field Gun 16 and Light Field Howitzer 16

This interim solution then led to the 7.7cm *Feldkanone* 16, the field gun of 1916. This received a longer tube with a different twist of rifling and still fired the same ammunition as the FK 96 n/A, but it no longer mated in one fixed cartridge, but separated. The shell and shrapnel remained the same, but the propellant was now in two different charges, the greater one containing more powder than before. This longer tube was mounted on the carriage of the new *leichte Feldhaubitze* 16, the light field howitzer of 1916, introduced at the same time. This also had a longer tube than the old lFH 98/09, and now a new breech too, with the Rheinmetall-*Schubkurbelverschluss,* already well proven in the FK 96 n/A, replacing the earlier Krupp-type *Leitwellverschluss* of lFH 98/09. Now it took one movement less to work the breechblock.

The lFH 16 also fired the same ammunition as the lFH 98/09, with a new shell introduced: the *C-Geschoss,* C-shell. The weights of FK 16 and lFH 16 were similar, with the thicker 105mm tube making up in thickness what the 77mm had in length. Rheinmetall did most of the design work on these guns, which were receiving in optics, laying mechanism, and so on, as many identical parts as possible. They were produced by several different manufacturers and the troops received them at the beginning of 1917.

At the end of the war German field artillery numbered 3,744 FK 96 n/A and 3,020 FK 16, together with 1,144 lFH 98/09 and 3,044 lFH 16. The new FK 16 caught up with the French 75mm and overtook the Russian 76.2mm. Like the FK 16, the lFH 16 also gave German artillery a lead against the enemies', both British 111mm and Russian 122mm field howitzers.

## The New German 15cm Guns of 1916

Germany had gone to war lacking in heavy flat trajectory long-range fire. Guns of 150mm (6in) were only standing at the coast and in fortifications. These old ring gun-type models of 1872 and 1892 vintage without recoil mechanism had to be accepted for the war of positions. We have already looked at the 15cm *Kanone in Schirmlafette,* the 6in gun with protective armour shield. This had been ordered by the APK in 1900 from Krupp, as a heavy gun able to change emplacements in the fortress cities of Strasbourg (which never received any) and Metz, where eight of the twelve made have been reported to have been

*The experiences of 1914 and 1915 finally led to the introduction of a new family of guns in 1916, mostly derived from the existing models and characterized by longer tubes. This started with the 7.7cm FK 16 L/35 (before L/27), which now rested on the carriage of the howitzer, thus allowing both greater elevation and range.*

117

*Developments of 1916 and Beyond*

emplaced. Transport was to be by railroad. Krupp designed it on the model of their naval guns with a short recoil system, firing 40kg (92lb) shells to a distance of 15km (9.4 miles) when the gun had maximum elevation of 30 degrees. It weighed 12,300kg (28,000lb), the tube alone 5,100kg (10,700lb). Unfortunately – from the German point of view – the chance to continue developing this gun, sitting with its pedestal on a railroad car, into a genuine railway gun, able to fire from the car without having to dismount first, was missed. Otherwise Germany would have had railway artillery before the war.

In 1908 transport from one emplacement to another was no longer limited to railways, but also attempted by road, by steam road locomotives, or even by horses, of which six were deemed sufficient. The new prime mover was a 60hp four-wheel-drive tractor built by Daimler at Berlin.

Later, the APK suggested increasing the range by firing a heavier shell of 50kg (110lb) with a lower muzzle velocity of about 700m/s (2,100ft/s). The heavy 5,100kg tube of this 15cm KiSL was the basis for designing a lighter one for a new 15cm gun, this time with long recoil and on a wheel carriage. Krupp tried to interest the APK, but they and the Ministry of War both saw no need for such a gun in 1908, the 15cm calibre already believed to be too heavy for good mobility. In 1914 they changed their mind and ordered one gun each for trials and comparison evaluation from Krupp and Rheinmetall. Since time was of the essence, both models were tried briefly in 1915 and then both were ordered rather than a decision being made in favour of one or the other. These guns reached the battlefield in 1917, and until the war's end Krupp supplied 214 and Rheinmetall 32. They were of similar design and differed mainly in their transport: the Krupp gun was designed to be pulled by motor tractors only and in one load; the

(Above) *The 15cm calibre was to be fitted with longer tubes. The experimental weapons of Rheinmetall for this venture included the 15cm Kanone 16 Rh.*

(Left) *The 15cm leichte Kartaune L/30 Rh, a mix of howitzer and cannon.*

*15cm* Kanone *16 RH, shown on pedrails firing* (top) *and about to push the tube into the carriage* (above).

(Below) *The competitor: the 15cm* Kanone *16 by Krupp, which was also introduced.*

*The 15cm K 16 Krupp: tube on cart* (top); *tube being pushed in from the front* (above) *and gun being winched forward* (right).

(Below) *The 15cm experimental howitzer L/30 by Rheinmetall.*

*Developments of 1916 and Beyond*

*An ex-naval 15cm SLK L/30 in a wheeled carriage* (left) *and a battery of them stopping on the march* (above). *The guns seem to be pulled by steam traction in one piece, despite their weight of over 10 tons with pedrails.*

Rheinmetall gun could be transported either in one load or broken into two and transported by horse.

## The New 10cm Guns

The 10cm *Kanone* 04, the 105mm (4in) gun of 1904, L/30, had also had a facelift, resulting first in the 10cm *Kanone* 14 L/35 and later in the 10cm *Kanone* 17 L/45, the range increasing from 10.2km (6.4 miles) with the old 04 to over 13.1km (8.2 miles) with the 14 and to 14.1km (8.8 miles) with the 17 model. During the war Krupp manufactured 724 guns of the 14 model and 192 of the later 17 model. The 14 was transported in one load, the 17 model in two. In spite of all of them, German long-range artillery kept falling behind that of the enemy in number. This was to be changed with the arrival of another makeshift weapon: the 21cm mortar of 1916.

## The New 21cm Mortar of 1916

The lengthening of the tubes continued to include the heavy 21cm mortar, which had been introduced after ten years' development by Rheinmetall and Krupp. They both received orders for manufacturing the mortar, of which the army had 216 when the war began. The need to increase range was also valid in 1916, as up to then the mortar had had to be emplaced close behind the front-lines because of its short range. This enabled early target acquisition by enemy artillery followed by a quick counterbattery bombardment. Therefore a longer tube (L/14.6 instead of L/12)

*Developments of 1916 and Beyond*

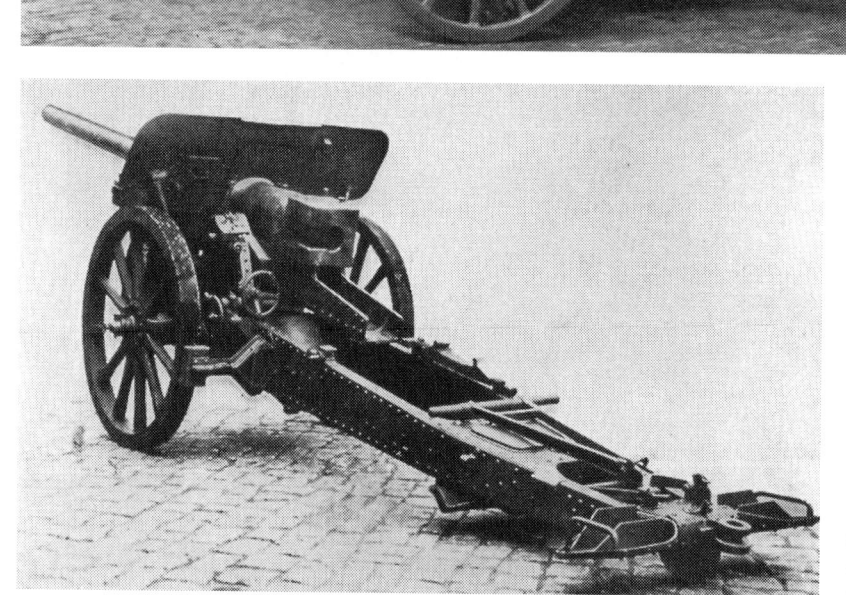

(Below) *The tube of the 10cm K 17 grew to L/45, but now it had to be transported in two loads.* (Inset) *Shown here is how the tube of the 10cm K 17 has been inserted from the tube cart standing behind the carriage.*

*The old 10cm K 04 was altered to the specifications of the K 17 and received the designation 10cm K 17/04.*

## Developments of 1916 and Beyond

was designed and mounted. The mortar was then called *langer* 21cm *Mörser* M 16, long 21cm mortar of 1916, and fired a 120kg (250lb) shell, the *Granate* 17, out to 10.2km (6.4 miles), instead of the earlier 9.4km (5.9 miles). This was not much of an improvement, but the aim had also been to keep the inevitable rise in weight as low as possible. This was successful, as the weight rose from 7,378kg to only 7,555kg (15,554–15,860lb). The first unsprung carriage was changed to springs so the mortar could later be moved by motor traction too, which unlike the gentle horses was a lot harder on the wheels, especially on the rough roads of those days. The mortar was moved in motorized traction in one load, with the tube drawn back.

### The 18.5cm Howitzer of 1917

Despite the fact that the troops were now satisfied with the long version of the 21cm mortar, it was decided to replace it with an 18.5cm (7.4in) howitzer. This was to combine a greater range of 11km (6.9 miles) with the lower weight of both howitzer (5,175kg/10,867lb) and shell (80kg/168lb). But this weapon did not fulfil expectations and was only built in a trial series of twelve howitzers.

*The 21cm mortar also received a longer tube of L/14.5 (before L/12) and was now called* langer Mörser, *long mortar, or 21cm* Mörser *16. It was still in use in World War Two.*

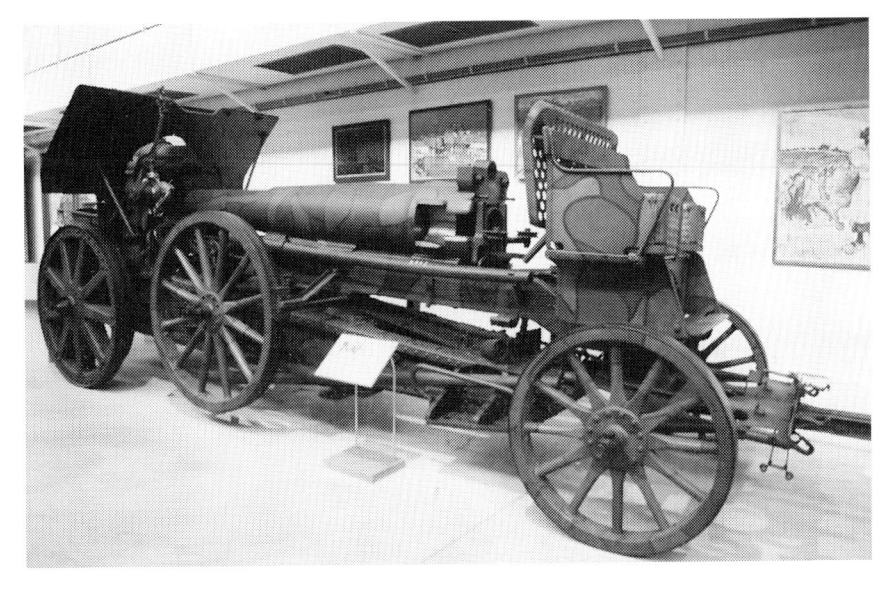

*The* lange Mörser *was rolling in two versions: this one was for horse traction and had the standard wooden artillery wheels.*

(Left) *The* gefederte *version of the* lange Mörser, *with springs over the steel wheels for faster motor traction, which quickly ruined the old wheels.*

*The new steel wheels* (inset) *and the trailer with the pedrails* (above).

*To close the gap between the 15cm and the 21cm calibre an experimental 18.5cm* Versuchshaubitze L/22 *was created by Krupp.*

124

## GERMAN RAILWAY ARTILLERY

The first heavy guns handed over by the German navy to its poor relation, the army, fired while bolted to the bottom of a round concrete bedding, the tube being moved in rough azimuth and elevation by winches, with the mechanism taking over for the fine laying. The 38cm (15in) gun was developed by Krupp within four weeks in September 1914 from their own arrangements on the Meppen range. This gave 24 degrees of elevation, which by tilting the bedding backward 5 degrees grew to 29 degrees and increased the range of firing. A more refined version then destined for the more or less surplus 35.5cm (14in) guns, an unusual calibre for the German navy, intended for the never built Mackensen-class (named after a soldier), then gave a greater range of a hitherto unheard of 62.2km (38.9 miles) at 52 degrees elevation. Here azimuth was at first taken with the help of winches, and elevation (or rather depression, the long naval tubes being front-heavy) was by pulleys. With the war of positions freezing the front-lines from the North Sea to the Swiss frontier, more long-range guns were needed to reach the enemy at places far behind his lines. On the other hand, guns of medium calibre also had to be used, as there were simply not enough of the real heavies, and in addition, using the big calibres sometimes proved ineffective as it produced overkills. This led to the further development of the 38cm *Bettungslafette*, the bedding mount, to permit an elevation of 45 degrees with better laying mechanism for azimuth and elevation.

The next step was at the end of 1915, with the first *Bettungslafette* for first the 21cm (8in) and then the 24cm (9.6in) SK. They went up to 45 degrees and had machines for taking the azimuth. And unlike the bedding of the 35cm and 38cm guns poured in concrete, these were of iron. And as a derrick was needed for the mounting of the gun anyway, these parts of the bedding could be few and heavy. This made them easy to transport, reusable and quick to construct. In fact, this system had so many advantages that it was also applied to the super-heavies of 35cm and 38cm. Derricks had to be in the range of the heaviest part to be lifted, normally the tubes. At first any crane or derrick was used, lifting between 30 and 150 tons. Later, special derricks were developed which could be taken apart themselves. When the crew got the knack, a single night was normally sufficient time to mount the gun ready to start firing.

Still efforts continued to do away with all these parts of the bedding and derrick. This led to the development of railway cars on which the guns were mounted so that they could fire directly from the car. The enemy was also known to have been working on this since the autumn of 1916. The French were reported to have used guns mounted on flat railroad cars inside besieged Paris in 1870–71. But they did not invent this themselves, and the credit must go to the US civil war of 1861–64. It seems that the confederate General Robert Lee had the idea in 1862, with the first gun firing the same year. It was a rifled muzzle-loader of 32lb calibre (17cm/7in). The Union countered two years later with a 13in mortar, which was baptized *Dictator*, as was traditional for big guns. And then of course there is also the alleged first inventor, a Russian named Lebedew, who is reported to have mounted a mortar on a railway car in 1860.

In World War One it seems that the French were a bit earlier to reanimate this old idea. It could have been Britain, where in 1900 it had already been suggested putting the almost new 6in field howitzer M 1 onto a genuine railway mounting. But when all the departments involved – from the Director General of Ordnance to the Superintendent of the Royal Carriage Department and the Inspector of Guns and Fortifications – had finished putting down their objections, it was 1902 and the latter found that it would be more practical to simply transport an ordinary howitzer on an ordinary car and unload it for firing when needed.

In France a Lt Col Peigné is credited in literature with having designed the first railway gun in 1883, but for me the later Commandant Mougin, famed for his designs of armour cupolas for fortifications and especially for his idea of the futuristic *fort de l'avenir*, is said to have put guns on railcars in 1870.

*Developments of 1916 and Beyond*

*The 17cm SLK had started rail transport for heavy guns, and larger calibres of naval origin followed. At first the rails only provided transport for these guns; they then had to be lifted onto beddings resting on large foundations. Later the guns fired immediately from the railway cars. This held for 24cm, 28cm and 38cm calibre. The 21cm Paris gun and the 35.5cm long-range gun stayed with the beddings. Here is the 24cm SK L/40, firing either from a bedding (bottom) or straight from the rails (top).*

In any event, the French, having huge numbers of old naval guns on hand which they, like the thrifty British Admiralty, fortunately hoarded after 1918, decided to put these on railcars. There were some designs that seemed impossible for any student of gun design to realize, such as the *affût à glissement*, the sliding mount, in which the entire car and gun and recoil were arranged to slide gently for a few metres along fat steel beams supporting them. But it worked.

So French railway guns improved, as did the Germans'. The second half of 1916 saw the first German 28cm and 38cm railway guns, the latter enormous with its eighteen axles. These were necessary because the guns had to roll without requiring special reinforced tracks. And the guns had to fit within the limits set by railway tunnels, for example, the elevation, which was 45 degrees with the 24cm and 28cm guns, was only 18 degrees at first for the 38cm. For firing, the cars were pushed by an engine into a curve until the gun was roughly pointing in the direction of the target; the gun car followed the ammunition car. Fine azimuth was then taken with the laying mechanism on the gun. Recoil was an astonishing 0.5–0.9m (2–3ft) only, thanks to brake shoes, with the 24cm and 28cm guns at low elevation; at high angle it was almost zero. The 38cm recoiled around 2.5m (8ft) and had to be pushed back by the engine

*Developments of 1916 and Beyond*

*The 24cm SK L/40 firing with the cars for ammunition* (right), *and crew still attached* (below).

(Bottom) *The 24cm SK L/40 firing from a turntable bedding.*

*Developments of 1916 and Beyond*

*The 28cm SK L/40 lifting the tube for firing. Camouflage does not seem to be have been too important on this day on the coast of Flanders. Maybe it is not* Flugwetter, *but* Fliegerwetter, *as the Luftwaffe had it, meaning the weather was no good for flying, only for the pilots now able to sit in the mess.*

into the correct position, the last inches being corrected by soldiers with crowbars.

This much movement only permitted slow firing, of course; it was certainly not quick enough to combat a ship sailing along. So firing from the coast was no longer done straight from the railway car, but with the gun lifted from it and lowered onto a firm iron bedding. A number of these were made ready at sites where an emplacement might be needed. The guns then received their own small derricks built onto the car, which assisted in lowering the smaller calibres onto the bedding, usual of the frontal pivoting type. The super-heavy guns of 28cm and 38cm needed other beddings. They were lowered onto a turntable by hydraulic winches and then the undercarriage pulled away. The gun could then fire inside a full circle, with an azimuth of 360 degrees, and could also take any elevation up to 45

*Developments of 1916 and Beyond*

degrees. Recoil was taken up by two recoil cylinders on the gun and in the end by a cylindrical abutment on the turntable. A recuperator pushed the tube with its counterweight of heavy steel plates on the breech forward again. These turntables also permitted the guns to be ready for action faster. A typical example of the emplacement for 21cm, 24cm and 28cm guns was that of the 28cm railway battery *Preussen*, one of the twelve built for three four-gun batteries on the Belgian coast. It was described in detail by two US officers, Majors Armstrong and Norton, examining it on behalf of the US Ordnance Department in December 1918. They wrote:

> The standard-gauge track extends a short distance beyond the emplacement, for the forward truck when removed from the mount… The auxiliary sections of track are laid inside the emplacement for the placing of the gun. As soon as the gun is in position, the central sections of rail… are removed and the base section of the pivot dropped and bolted in place. The carriage is then raised a few inches by means of jacks… and the trucks removed. The short sections of track within the emplacement are likewise removed and the rear of the carriage let down until the traversing rollers rest on the roller path.

This method of emplacement was also used on the heavies without a turntable, up to the heaviest of them all, the 410-ton 21cm *Parisgeschütz*, Paris gun, L/162.

The US authors also stated that 'the range of 55km (34.6 miles) reported for the Deutschland battery seems extreme'. (Remember the 35.5cm (14in) L/52.6 guns named *König Wilhelm*, King Wilhelm, of the planned Mackensen-*Klasse* which ranged out to 62.2km/38.9 miles? And this not at the cost of short life expectancy: one 35.5cm tube fired 578 rounds without damage or resulting wear. That was Krupp's steel.) The duo also looked for

*The 28cm SK L/40 could either fire from a bedding* (below) *or directly from the rails* (top).

*The 38cm SK L/45 was the largest calibre of the railguns in World War One. It is shown here on a bedding with an azimuth of 360 degrees.*

*(Below left) Firing started with the ammunition being rolled on carts to the gun, which has the tube already in the loading position of 0 degrees. The shell weighs 1 ton.*

*The shell truck is pushed towards the open breech end of the tube. Then the rammer lying right of the gun will be manned (with 38cm calibre it took eight to twelve men to seat the shell properly) so the lands of the tube (the raised parts) bite into the copper driving bands of the shell, hard enough to hold this even when – after the propellant has also been loaded and the breechblock closed – elevating the gun would not let the shell fall back onto the powder. Since the diameter of the chamber is always larger than the bore, this would have let the gases developed by the powder pass by the shell, resulting in a short-falling round of less than 50 per cent of the calculated range.*

*Then the gun is elevated to the degree prescribed by the firing table according to the distance of the target. The heavy steel counterweights on the tube serve as equilibrators. This was made necessary by the trick of increasing range by installing the jacket of the tube the other way in the mounting, so the trunnions, not lying in the middle of it, came further back and the gun could elevate higher. The price was more counterweight.*

*Developments of 1916 and Beyond*

which was supposed to have fired mostly against the places of Ypern and Dünkirchen. It had housed a single 38cm Krupp-gun, model 1914 L/45. They found it well protected, though it only had an azimuth of 157 degrees.

> The emplacement for this carriage is of extremely massive concrete construction. The diameter of the central pit or well in which the carriage is placed is approximately 22.439m (67.5ft)... The depth to the level of the traversing rack is 3m (9ft) and the additional depth to the floor on which the center pintle rests is 1.5m (4.5ft), making a total depth of 4.5m (13.5ft). On either side are practically identical concrete structures for the housing of ammunition and personnel... The thickness of the roof of the structures on the right and left is 3m (9ft), the total height above the floor being 5.5m (16.5ft).

A similar arrangement was found in the 35.5cm (14in) batteries. They also examined the armour which enclosed the Pommern battery and reported:

> To protect the personnel operating this gun against aircraft bombs and aircraft machine gun fire, the carriage was covered with 6cm (2.4in) flat armour. This plating extended to within a few centimetres of the floor of the pit. The hole in the front through which the guns extend is sealed by the small shield on front of the cradle.

*And now the gun fires. The third soldier not holding his ears must already be deaf.*

hits by their own side made on the guns and found none, reporting,

> The writer saw many of the holes made by airplane bombs... but could not find a single case in which either the inland batteries or the batteries in the dunes had been struck by shell firing from the sea or by bombs dropped by airplanes.

This is explained on the part of the British monitors by the dense smoke screens, and on the part of the pilots by the dense and accurate antiaircraft fire. They also visited the 38cm Pommern battery at Leugenboom, somewhat to the rear of the beach,

Obviously these were not ship turrets, but the famous rolled homogeneous armour (RHA), which today is the yardstick by which to measure and compare the penetration power of shaped charge warheads. The thickness was good enough against machine gun bullets and ordinary shrapnel or splinters. Bombs would have been another story, depending on their weight. But as discussed, dropping bombs and hitting the target are two different things, as is firing and hitting the target. This was good for the crew, and was commented on as follows:

> Gen Arnoulde's report [of course, the French had also understandably stuck their noses into the

*Developments of 1916 and Beyond*

German guns and their positions – who would not?] states that the projectile was rammed by 12 men and the rate of fire was one shot in five minutes with electrical operation of the elevating mechanism, and one shot in 10 minutes with hand operation of the elevating mechanism. Gen Arnoulde's report further states that according to reports of people living at Leugenboom, the personnel originally provided for the operation of the mount when it was operated by hand was 1 captain, 2 lieutenants, 10 noncommissioned officers, and 160 men. After provision was made for electrical operation of the various mechanisms, the personnel was reduced to 1 captain, 2 lieutenants, 5 or 6 noncommissioned officers, and 70 men.

This seems to point to the missing ninety men having been engaged in cranking handwheels. It is of interest to compare the eighty men of the mechanized crew of this single 38cm coastal gun with the contemporary crew of a British 15in twin turret of the Vanguard class. The number was almost the same – eighty-three men with the officer in charge – but they operated two guns of the same calibre. Since I do not believe the British sailor of those days to have been twice as strong as his German counterpart (coastal artillery had been the business of the German navy since 1888), the explanation must be found in the technical arrangements for ammunition supply, with the onboard lifts working both faster and without needing muscles, compared to the trolley pushers.

The railguns, practically the next stage in development of the super-heavy low-angle fire already discussed, started at 15cm calibre and went up to the 38cm. The guns kept the same first names they had held in their makeshift beddings. One exception was the 17cm (6.8in) *Samuel*, because he started his career as a ship gun, then received a plump carriage with wheels to give him mobility on land, and was then named 17cm SK L/40 *in Räderlafette*; its 23.5 tons too heavy for horses and even the weak tractors of this time. Later, thirty of these guns ended up as railguns, standing on a flat railway car, with the carriage wheels looking quite superfluous.

The 21cm SK L/40 and L/45, together with the 24cm SK L/30, L/35 and L/40 guns, were no longer mounted on normal railway cars, but on specially built carriages, the 80-ton *Staatsbahnwagen* of the national railways, with the aforementioned two winches on each side. Their four axles had to be raised to five for the next calibre, the 28cm SK L/40 in *Eisenbahn-Bettungs-Schiessgerüst*, railway-bedding-firing rack, as the constructions were called. In 1917 this was finally topped by the heaviest calibre of all German railway guns: the 38cm SK L/45, called Max.

The limited elevation of big guns on railway cars, caused by the breech end threatening to dig into the ground when firing on high elevation, was met by a simple change in the 28cm and 30.5cm guns. This involved simply turning the cradle containing the tube around, so that the trunnions, formerly in a forward position, were now at the rear. This gave an elevation of 55 degrees, even when firing from the tracks. The price for the extension of the range was only a pneumatic equilibrator. The guns were christened Lützow, and this method was applied to both the 28cm SK L/45 and the 30.5cm SK L/50. A similar change was planned for another 38cm SK L/45 gun, which was to be named Wotan.

Of course, there were also captured railway guns. The Russians' 152mm rail howitzers were not used by the German army, but three British 12in railway guns were.

Germany entered World War Two with a lot of the old railway guns interned after 1918 until 1935, and then built an additional few. These included the super-heavy 80cm *Kanone* (E), 32in railway gun (E = *Eisenbahn*, railway), writing about which has fed whole generations of military authors' families. Another was the K 5 (E), the *Kanone* 5 (railway). In this case, as with other railway guns, the number after the K stood for the range in tens of kilometres, thus '5' indicates that this gun fired 50km (31.2 miles).

By the end of World War Two, railway artillery was finished. In Germany it could no longer move on the destroyed tracks, and unfortunately the heavy bombers had taken over on the other side.

132

## Railway Artillery

| Gun model | Calibre (in) | Weight Empl. (tons) | Tube Length (in) | Shell Weight (lb) | Muzzle Velocity (ft/sec) | Max. Range (ft) | Elevation/ Azimuth (degr.) | Remarks |
|---|---|---|---|---|---|---|---|---|
| 15cm L/45 | 5.9 | 55.5 | 270 | 92.4 | 2,520 | 68,100 | +45/180 | (1) Krupp |
| 17cm L/40 | 6.8 | 129.1 | 272 | 134 | 2,445 | 72,000 | +45/27 | (2) Krupp |
| 21cm L/45 | 8.4 | 108 | 378 | 262.5 | 2,670 | 79,200 | +45/2 | (3) Krupp |
| 24cm L/30 | 9.6 | 113 | 288 | 312 | 1,920 | 56,100 | +n/a /360 | Krupp |
| 24cm L/40 | 9.6 | 109 | 384 | 399 | 2,520 | 76,740 | +n/a /360 | Krupp |
| 28cm L/40 | 11.2 | 166 | 448 | 603 | 2,490 | 96,000 | +45/360 | (4) Krupp |
| 30.5cm L/18.8 | 12 | 61.7 | 226 | 714 | 1,341 | 39,300 | +45/120 | (5) Elswick |
| 35.5cm L/53 | 14.2 | 240 | 752.6 | 722 | 3,495 | 186,600 | +52/ | (6) Krupp |
| 38cm L/45 | 15 | 268 | 1,575 | 2,240 | 3,120 | 142,500 | +55/360 | (7) Krupp |

All German railguns were former naval guns, firing either after dismounting onto a bedding or later from the railcar itself; all bore names.

*Remarks:* (1) gun inside armour box, twenty-four made; (2) naval guns first put on wheels: too heavy; then on railway, thirty made; (3) transferred from bedding onto rails, only two made; (4) seven built; (5) ex-British 12in howitzer Mk 3, three captured; (6) only transport by rail, firing from bedding, only one gun made; (7) only one made.

n/a = no data available.

This was not even changed when a fantastic new concept appeared: the first cruise missile, then launched by the German airforce and named *Vergeltungswaffe* 1 (V1), reprisal weapon 1, by Hitler. Neither this bomb flying on a preplanned course, nor the army's rival *Vergeltungswaffe* 2 (V2) was able to change the outcome of the war. After the war the German scientists who had developed it helped both sides to build up an overkill of these liquid fuel rockets, which have already taken mankind to the moon, and could ultimately lift it to the stars or destroy it completely.

## INFANTRY GUNS

Infantry guns were not an invention of World War One. The idea had already been employed in earlier wars. Both the light 'leather guns' of the Swedish army of Gustav Adolf and the *Bataillonsgeschütze*, the battalion guns of Frederic the Great, fell into this category, as did the light field guns of the troops of the French revolutionary army, later returned first to the artillery by Napoleon I and then back to the infantry. After this time the guns of the artillery were judged light enough to accompany the infantry. Another school even advocated the idea that the field artillery could in most cases support the infantry by firing out of its emplacement. The war in Africa of 1900–1902 was an exception to this, with both the British army and Boer troops fielding small-calibre guns along with their infantry.

No army had anything like a true infantry gun before World War One, however. The topic had occasionally been discussed in military literature, but no definitive answer given. Some writers argued for small, light guns, keeping in close contact with the advancing infantry. This was reasoned to be caused by the inability of field artillery to reconnoitre and neutralize objects such as machine guns covered by woods or in houses. But the promoters

*Developments of 1916 and Beyond*

of the field gun, being in overwhelming numbers, were of course right. And then there was the sensible argument that increasing the guns of the light field artillery by introducing a third type would be simply too much.

The war revealed who had been right. The German field artillery was no longer able to accompany the infantry on the battlefield, especially when in later years the quality of horses declined. The FK 96 n/A was too big a target, even over long distances; it drew the enemy fire and was knocked out in its open gun position within a short time. In soft terrain such as at Armentières, the guns could only be brought forward with the aid of *Armierungssoldaten*, half military labourers in uniform, whose duty it was to work on the *Armierung*, the building of fieldworks around a fortress, planned in peacetime and realized at the beginning of a war, using materials already stored for this purpose. At other places the guns were stopped by ditches or steep heights, or they were discovered by the enemy and immediately neutralized. The same was especially true of the war in the east, where the soft ground later became famous for its *rasputiza*, roadlessness, in spring and autumn. At a meeting held in Berlin, some troops from the east were all for arming the infantry with a light gun of its own, whereas the majority from both west and east still believed the FK 96 n/A to be light enough to accompany the attack of the infantry.

At the beginning of the war the lack of cooperation between artillery and infantry caused unnecessarily high losses. The infantry sometimes believed that they could do without artillery, and communication was bad anyway due to the lack of telephone and radio equipment. The field artillery in their covered positions simply could not locate targets that were dangerous to the infantry, never mind fire onto them. When the war of positions came, the discussion about the infantry gun was silenced, as by then the infantry had received three weapons fit for this purpose: the 7.7cm *leichte Minenwerfer*, the 3.7cm *Grabenkanone*, trench gun, and the 3.7cm *Sturmbegleitkanone*, assault accompanying gun. All were well suited to the trench warfare.

## The 7.7cm *Leichte Minenwerfer*

We have already encountered the 7.7cm *leichte Minenwerfer*, light mine launcher, together with its two heavier brothers, as the poor man's artillery of the engineers, and later of the infantry. They had been developed by Rheinmetall at the suggestion of the *Ingenieurkomitee*, the engineer's committee. The heavy 25cm MW dated from 1908, and in 1911 the medium 17cm MW was ready. In 1909 Rheinmetall then received the order to design the 7.7cm light gun, which was tested in 1914 before the war started. Like the other two it was a rifled muzzle-loader. The breech end cap screwed into the tube and contained the spring-loaded firing pin. (The other two, of heavy and medium calibre, were fired by electricity.) All three were built like modern guns, with hydraulic recoil cylinders and spring recuperators. The sighting and laying mechanisms were also those of the artillery. The light MW, which would normally fire in the high-angle mode like its brothers, could also fire with low trajectories. For this a tail could be fixed to its mount. In this configuration it served as an antitank gun.

The important data was L/5.2, resulting in a tube of 400mm (16in) in length, firing between 0 degrees and +27 degrees in the low-angle mode and +45 degrees and +75 degrees in the high-angle role. It ranged between 160m and 1,300m (480–3,900ft) according to the muzzle velocity, which could be increased from 77–121m/s (231–363ft/s). The weight emplaced was 140kg (295lb) for the lMW alone, and 215kg (450lb) with the tail added. Because of its ease of manufacture, weight and handling, it had replaced the field gun to a certain degree, especially when firing the new shells it had available now as well as the explosive *Sprengmine*. These were smoke-, gas-, illuminating-, information- and antitank-mines/shells.

## The 3.7cm *Grabenkanone*

The 3.7cm *Grabenkanone*, the trench gun, was a makeshift design by Krupp, destined for the front-line of the trenches. It was thought to supplement the light mine launchers and to neutralize indirect fire targets such as machine guns, armour shields, cover and blockhouses. It was an offspring of the old five-barrelled Gruson revolving guns of navy and fortifications alike, becoming superfluous as a ditch defence gun in 1890, when it was replaced by the single shot quick-fire 5cm casemate gun. Now each one produced no less than five trench guns, by having each of their five tubes fitted with a bayonet-joint breech. It was mounted on a simple construction riveted together from sheet steel and U-shaped beams with two different armour shields protecting the crew from enemy hand-gun fire and fragments. A small gun was used to fire from an enclosed position and a large one for firing from an open position. The mount of the trench gun was fixed to the ground by hammering two spikes into it, these being sufficient to take up the recoil. The iron sights were graded from 0 to 1,400m (4,200ft); the gunner could observe them with the help of a mirror from the side, so that he did not have to expose himself to enemy fire.

The trench gun fired from its L/21.5 tube a shell of 455g (1lb) with 40g (0.09lb) explosive inside, detonated by an impact fuse. The muzzle velocity was 400m/s (1,200ft/s) and the range up to 1,500m (4,500ft). The weight of tube with elevating mechanism was 38kg (80lb), the bayonet-joint breech with recocking firing pin another 3kg (6.3lb). Over short distances the 168kg (360lb) of the emplaced trench gun could be broken down into the three loads of tube (38.3kg/86lb), mount (125kg/263lb) and the 13mm (0.5in) thick armour shield used according to the emplacement (25kg (53lb) for the small and 45kg (95lb) for the large one) and then carried by the crew. For longer distances it was loaded onto a car.

## The 3.7cm *Sturmbegleitkanone*

The 3.7cm *Sturmbegleitkanone*, the assault accompanying gun, was also a Krupp design. It had been devised in 1915 as a weapon to accompany the infantry in an attack out of their own trench toward enemy lines. It too was based on the tubes of the 3.7cm revolving gun and may be considered the forerunner of the infantry gun. The tube was one of the five of the Gruson-Hotchkiss revolving gun and the breech end was also closed by a bayonet joint

*The need for the infantry to have guns of their own resulted, as a first solution, in the 3.7cm stationary* Grabenkanone, *trench gun, which Krupp made using the tubes of the old Gruson 3.7cm revolving guns, retired from the fortifications to the depots by the machine gun.*

*Developments of 1916 and Beyond*

with a recocking firing pin, as in the trench gun. This tube was mounted between two parallel sheet steel walls, bolted together at a certain distance, with a firm spike fixed at the rear and an axle for two steel wheels in front. The L/21.5 tube was mounted in a fork which was able to be turned 10 degrees to each side. The handwheel of the elevating mechanism raised it between −10 degrees and +60 degrees.

The armour shield was in four parts: middle (6mm/0.25in), lower and two side shields (each 5mm/0.2in). The ammunition was carried on the trail of the gun in a box holding twenty 3.7cm cartridges. The trail also held two planks, used as a bridge when crossing ditches and fixed to the trail when not needed. Sighting was done by telescope, with iron sights as a reserve. As the gun could fire in both modes – direct and indirect – a gun director was included. Azimuth was 10 degrees to each side and elevation ample: −10 degrees to +60 degrees. The gun fired three sorts of projectiles: impact fused shells against troops, armour shields and blockhouses; canister with forty-eight balls of 13g (0.5oz) each; and bombs launched from the trench in the high-angle mode into the enemy trench. These bombs had an impact fuse and held 1.42kg (3lb) of explosive, with another 470g (about 1lb) of explosive in a hollow shaft, adding up to 1.89kg (4lb) altogether. The hollow shaft was of less diameter than the calibre, so that it could be inserted into the tube, transmitting the push of the propellant. This was made up of five different charges, resulting in a muzzle velocity of 400m/s (1,200ft/s) for the shell and ranging from 33m/s (99ft/s) to 80m/s (240ft/s) for the bomb. For the canister, the range varied from 200m (600ft); for the bomb up to 400m (1,200ft); and for the shell up to 3,000m (9,000ft). The gun weighed 370kg (780lb) in its emplacement.

For transport it was placed on the captured French undercarriages of the 9cm gun, which weighed a lot more than the gun itself: 615kg (1,290lb). Over short distances and when accompanying the infantry during attacks, the gun was pushed by its crew at the crossbar; they were well protected behind and inside the surrounding armour. The width (120mm/4.8in) and diameter (630mm/25in) of the wheels were suitable for crossing the craters of the battlefield.

Though it seemed that, at least for the time being, the problem of the lack of infantry guns had been solved by these three weapons, the Ministry of War continued work on this problem. It seemed especially important for the theatre of war in the east. As there were no experiences to learn from and the gun-making industry did not have the capacity to develop a completely new design, other solutions were needed. One of them was to use captured guns of suitable weight and calibre, if they existed. They did.

### 7.62cm Putilov Fortress Gun

A huge number of close defence 7.62cm Putilov fortress guns, 3in guns of 1910, had been found in the Russian forts that had been taken by the Germans. They had already been tested on the western front (it is to be presumed that this was in a slightly modified form, their more or less fixed fortress pivot mounts being exchanged for a more mobile wheeled carriage) with the new *Sturmabteilung*, the battalion of attack. Now Krupp was ordered to redesign them for the role of an infantry gun. This happened with the first forty-eight of the 7.62cm *Infanteriegeschütz* L/16.5 arriving at the front in the summer of 1916.

### 7.62cm *Infanteriegeschütz* L/16.5

These guns also went to the new battalions formed for attacking. The design had kept the tube and the screw-type breech of the original guns resting on a box-type carriage, with seats for the crew. These were not the old comfortable bucket seats already mounted on the guns of C/64 and C/73 vintage, on which two gunners could ride along facing the rear, but two rather simple narrow seats both facing towards the breech, so the gunners could do their job in a lower and therefore safer position behind the rather small shield. The gun now weighed 608kg (1,270lb), comparing well with the 1,020kg

(2,140lb) of the heavier FK 96 n/A, when emplaced, and fired captured Russian ammunition: canister up to 600m (1,800ft). These were presumably filled with heavy balls, since the maximum distance over which canister was effective on human targets had been considered by the German army to be 300–400m (900–1,200ft) at the most, because the balls lost so much kinetic energy during the flight towards the target. The shells were especially manufactured for this gun by Rheinmetall. The sighting equipment worked up to 3,000m (9,000ft), and for longer ranges a goniometer had to be used.

At first the troops liked everything about this gun: weight, accuracy and the effect of the shell. Later it was found that the gun had a low life expectancy due to the poor quality of the Russian steel used in its making. When the tubes wore out, accuracy also suffered. And it could fire only in the low-angle mode, with its elevation running between +11.5 degrees and an astonishing –18.6 degrees. This was a legacy from its fortress ancestry, when it had been designed to fire from embrasures high in the walls down into the ditches of a fort. For any range over 2.7km (1.7 miles) the tail of this gun had to be dug in the familiar way. Krupp therefore designed a new gun for the same purpose.

## The 7.7cm *Infanteriegeschütz* L/20 Krupp

By the autumn of 1916 this gun had already reached the front, only a few months after the L/16.5. This incredibly short development time was only possible because Krupp reverted to using many well proven and already existing components. These included the shortened tube of the 7.7cm FK 96 n/A, mounted in a carriage of a Krupp mountain howitzer. This had the advantage of having been designed to break down into several loads already. The gun fired the ammunition of the FK 96 n/A, but as a rule with a reduced charge only, which lowered muzzle velocity from the normal 435m/s (1,300ft/s) to the lower 400m/s (1,200ft/s), but if need be the normal charge could also be fired. This need arrived in the form of enemy tanks, when the gun fired a new armour-piercing shell with the full old power. Having taken a close look at the moon-like cratered terrain of the battlefield, the weight had been set at only 500kg (1,050lb), as the crew was supposed to pull it there cross-country. So the gun was moved in two loads, with the tube on a cart of its own as well as the mount on its wheels. On the road four horses were to do the same job in one load. Over really heavy terrain the gun could be broken down into no less than eight loads.

The soldiers were content: the gun fired a shell of 6.85kg (14.6lb) up to 5,000m (15,000ft), and the tube also elevated satisfyingly from –7 degrees up to +30 degrees, sufficient for the present. But there was a price to pay: the gun weighed 855kg (1,900lb) when emplaced, and breaking down the gun for transport and putting it together again for firing took time.

When in 1916 the new models of field guns 16 and field howitzers 16 arrived, they were more effective, but they also had a much higher weight. By this time the first tanks had made their appearance on the battlefield, so it was thought time to give the German infantry a light gun of its own. With ideas on how it should look still rather vague, it was time for another interim solution, especially as parts from the 1905 FK 96 n/A were available now that its production was running out. This led to the 7.7cm *Infanteriegeschütz* L/27 Krupp.

## 7.7cm *Infanteriegeschütz* L/27 Krupp

This was built mostly out of parts of the field gun 96 n/A, for example tube with breech, recoil mechanism and carriage. Changes included smaller wheels set closer together and the omission of seats and the lower part of the shield. Still the resulting weight of 845kg (1,780lb) was too high for transport on the battlefield. So like the L/20, the gun was broken down into two loads, but no longer also into eight loads.

The ballistics were more than sufficient. The army had asked for a combat range of only up to 2,500m (7,500ft), but they got a lot more: a range of 4,600m (13,800ft), and even over 7,800m

(23,400ft) with the old process of digging in the tail. But muzzle velocity was too high for undulating terrain, resulting in too much dead ground, and elevation went from −15 degrees to a mere +12 degrees. The army had to be content with this and received eighteen batteries of these guns in spring 1917. But the High Command was not happy and declined the request for their production in quantity. They went to look elsewhere. Their first step was to contact the three gun manufacturers available: Krupp, Rheinmetall and Austrian Skoda, so they could draw on their wartime experiences to create an ideal infantry gun. Rheinmetall finished its homework first and in the spring of 1918 revealed their 7.7cm *Infanteriegeschütz* L/19.5.

## 7.7cm *Infanteriegeschütz* L/19.5

This gun also used both the tube – shortened by 7.5 calibres – and breech of the FK 96 n/A. But it received a new recoil mechanism and a longer carriage to give it more stability during firing. This would have hindered mobility, however, and was therefore made to fold. Even in this position firing was possible with the aid of small spades stuck in the swivel of the tail. It used the same ammunition as the L/20 and L/27: as a rule the reduced charge of 400m/s, in exceptions with the full one of 435m/s. Its normal range was over 4,500m (13,500ft), or in special cases 5,100m (15,300ft). Weighing a high 802–817kg (1,700lb) depending on the carriage, it had to be transported in two loads, with tube and armour shield lying on a cart of their own. If the crew had to pull it their toil was hopefully made easier by an extra wheel under the tail. This was another makeshift solution with all the drawbacks of such, and only one battery was ordered.

The experiences gathered with smaller calibres such as the captured French 37mm (1.5in) infantry gun, model 1916 T.R. (*tir rapide*, rapid fire) and the Austrian 37mm M. 15 *Infanteriegeschütz* had not been satisfying either. And when in the spring of 1918 the OHL needed a lot of infantry guns because it had decided to arm each infantry regiment with a six-gun battery of their own, they did not have the necessary infantry guns for this. Only fifty of Krupp's unsatisfying L/27 guns were available. So the Germans borrowed the 7.5cm M. 15 *Gebirgskanone*, the 3in mountain gun model of 1915, from the Austrians.

## 7.5cm M. 15 *Gebirgskanone*

In spite of Austrian pride in this gun, the Germans were still not satisfied with it as an infantry gun. The wheels were too small for the shell holes, and who wanted a gun which could be broken down into many loads as an infantry gun? But Germany already had enough of these to equip fourteen batteries, and they ordered enough for another sixty.

The spring offensive of 1918 taught a number of lessons about the type of gun necessary for the infantry. Afterwards the following guidelines were laid down:

- the mass of the gun should not exceed 500–600kg (1,100–1,320lb), so it could be pulled by the crew across the battlefield in one load
- the height of the trunnions should be 600mm (24in) at the most, with an armour shield against machine gun fire from 400m (1,200ft)
- the range should be up to 2,500m (7,500ft); it should have the accuracy of a field gun; muzzle velocity should be at least 350–400m/s (1,050–1,200ft/s)
- a single shell should have a high effect, therefore calibre should not be under 7.5cm (3in); not many different shell types should be used
- it should have a high rate of firing
- provision should be made for the supply of sufficient ammunition, even during a battle
- future experience may result in further demands.

Krupp and Rheinmetall were both asked to submit designs based on these guidelines as soon as possible. They could even work on one based on a 5cm (2in) calibre gun, contrary to the rules above.

Krupp finished its gun quickly, and the *Infanteriegeschütz* 18 was just as quickly accepted by the Ministry of War.

*Developments of 1916 and Beyond*

*5.7cm ex-Belgian Nordenfeldt-fortress gun, employed in the trenches as an infantry gun. Note the fixed cartridge ammunition permitting a high rate of fire.*

### *Infanteriegeschütz* 18

This again used the ammunition of the 7.7cm FK 96 n/A, but with a lower charge, leading to a 350m/s (1,050ft/s) muzzle velocity and a range of 5,000m (15,000ft). The pressure of time had resulted in Krupp using a shortened version of the FK 96 n/A tube as well as its carriage, but now with an additional folding tail. The gun weighed 650kg (1,370lb) and was transported by horses on the road, and for short distances on the battlefield by the crew. Elevation was only to +15 degrees, limiting the gun to low-angle fire. The cartridges again held no less than four different shell types.

A large number of these guns were ordered at once, with some finished by the autumn, but not reaching the troops. Later the *Reichswehr* took them over.

Rheinmetall then submitted their designs: the 7.5cm *Infanteriegeschütz* L/20 and the 5.7cm *Infanteriegeschütz* L/30.

### 7.5cm *Infanteriegeschütz* L/20 and the 5.7cm *Infanteriegeschütz* L/30

The novelty of these guns was their semiautomatic breech system, which increased the rate of fire a good deal. The cartridge was rammed into the open breech, forcing back the extractors which held it open, and the breechblock to move up. After firing and end of recoil, when the tube moved forward again and gas pressure had dropped to a safe low, the breechblock was cammed down again, with the extractor first levering out the empty cartridge case and then arresting the breechblock in the open position. The cycle of loading-firing-unloading could then start again. This design has lasted through the tank guns of World War Two to modern artillery, though cartridge cases are rarely used.

In order to test these two models, eight guns were ordered of each, but they were not finished by the end of the war.

Besides these guns, the army also ordered twelve guns of another Austrian design for tests: the 7.5cm M. 17 *Nahkampfgeschütz*, the 3in close-range fighting gun of 1917 made by Skoda.

### 7.5cm M. 17 *Nahkampfgeschütz*

Skoda had started work in 1916 when asked to by the German authorities, and finished the gun one year later. But the order came too late for any of these to be finished by the end of the war. The gun was derived from the famous 75mm (3in) Skoda mountain gun of 1915, with the tube shortened from L/15 to L/12 and with a muzzle velocity of only 190m/s (570ft/s). The shells weighed 6.5kg (13lb) and covered a range of up to 3,000m (9,000ft), thanks to the carriage giving an elevation of between –10 degrees

*Developments of 1916 and Beyond*

and +70 degrees, the latter stemming from the mountain-gun heritage. The same was true for the transport of the gun, which was possible in three different ways: drawn by one horse, carried by pack animals in four loads, or by men in eleven loads. Looking much like the infantry guns of World War Two, it might have been just what was necessary for the infantry in World War One.

In World War One, the fifty-one batteries of infantry guns, each with four guns, demonstrated one thing: asking for a multipurpose weapon, able to fire both in high and low angle, resulted in a heavy gun which was a conspicuous target on the battlefield. And even though the German army had voted against such a combination of the *Minenwerfer* and the *Feldkanone*, it accepted these for a time.

In the end it was realized that a special weapon was needed for combating the enemy tanks. This led to plans to arm each infantry battalion with two infantry howitzers and also two to four antitank guns of 37mm; they were plans that were never realized.

This was undertaken after 1935 by the *Wehrmacht*, however, who next to their first antitank gun, the well-known 3.7cm *Panzerabwehrkanone* (*Panzer*, armour, is an abbreviation of *Panzerkampfwagen*, armoured fighting car) designed by Rheinmetall in the 1920s and sold worldwide, also introduced *Infanteriegeschütze*, infantry guns. The lighter one introduced in 1927 was the 7.5cm *Infanteriegeschütz* 18; but this was no longer the Krupp design of World War One. It was both designed and built by Rheinmetall and had a unique tilting tube. The calibre then went up to the heavy 15cm *schwere Infanteriegeschütz* 33.

Today the infantry can drag with them all sorts of launchers, rockets and rifle grenades, as well as calling for supporting artillery fire or a thunderbolt from airplanes. But they have also started to request guns of their own again, and they have got them quickly, most of them in the 30–40mm calibre range and firing long bursts from full automatic machine cannon, which is what they are, even

### Infantry Guns, Trench Guns and Assault Guns

| Gun model | Calibre (in) | Weight Empl. (lb) | Tube Length (in) | Shell Weight (lb) | Muzzle Velocity (ft/sec) | Max. Range (ft) | Elevation/ Azimuth (degr.) | Remarks |
|---|---|---|---|---|---|---|---|---|
| 3.7cm Graben-K | 1.5 | 710 | 31.8 | 1 | 1,200 | 4,500 | +8/22.5 | (1) (2) Krupp |
| 3.7cm SturmbegleitK | 1.5 | 777 | 31.8 | 1 | 1,200 | 9,000 | +60/10 | (1) (2) (3) Krupp |
| 7.85cm IMW | 3.1 | 452 | 16 | 9.7 | 363 | 3,600 | +27/7 | (3) (4) Rheinm |
| 7.62cm IG L/16.5(R) | 3 | 1,277 | 50.8 | 12.6 | 885 | 12,000 | +11.5/9.5 | (1) (3) (5) Putilov |
| 7.7cm IG L/20 | 3 | 1,796 | 61.6 | 14.4 | 1,305 | 15,000 | +30/5.5 | (1) (3) Krupp |
| 7.7cm IG L/27 | 3 | 1,775 | 83.2 | 14.4 | 1,305 | 23,400 | +12/6 | (1) (3) Krupp |
| 7.7cm IG L/19.5 | 3 | 1,684 | 60 | 14.4 | 1,305 | 15,300 | +15/5 | (1) (3) Rheinm |
| IG 18 | 3 | 1,365 | 71 | 14.4 | 1,050 | 15,000 | +15/7.5 | (1) (3)( 6) Krupp |

*Remarks:* (1) firing cartridges; (2) tube of Gruson revolving fortress gun, fixed position; (3) wheeled carriage; (4) firing with extra tail in flat trajectory mode; (5) ex-Russian fortress gun; (6) ammunition of FK 96 n/A with less propellant, folding tail.

though they are named *Abschussgerät*, launching device, and sport the handlebars of a big Harley Davidson, just like the H&K version.

## ANTITANK GUNS

If the age-old question: which was here first, the chicken or the egg? was asked of the tank or the antitank gun, there is no doubt: the antidote came first. It need not have been this way, as we shall learn in Chapter 4. But in World War One, for the Germans the antitank artillery came before the tanks.

The signs had been there to see since the Battle of the Somme, from 1 July to 13 November 1916. This had seen the first tank attacks by the British. In the end the Allied offensive failed. What had been the reason? The German High Command wrongly understood it to have been the success of the German antitank actions. Later, the authorities saw it differently: the forty-nine British Mark I tanks had only arrived on 30 August. Their attack had been ordered for 15 September, despite strong objections by both Churchill and Joffre. In the meantime, the high British and French expectations for this new weapon led to continuous performances of them in front of high-ranking officers, exhausting the crews before the attack. They therefore went into battle fatigued.

The tanks were launched in groups of two to three against certain targets, which they were expected to reach five minutes before the infantry. Only thirty-two of the forty-nine tanks reached the starting positions. Then nine broke down there and another nine were not able to start in time, but did so only later. Of the fourteen tanks departing and attacking as planned, five (37.5 per cent) were hit by the German artillery. The remaining nine were able to make a breakthrough into the German front of an average depth of 3km (2 miles).

Further smaller tank actions followed in 1916. The 'Easter Battle' at Arras brought another disaster for the tanks, with only thirty-four out of a planned sixty starting on 9 April; seventeen (50 per cent) were hit by the artillery and some of them captured.

The new 8mm *Spitzgeschoss mit (Stahl-)Kern* (SmK) bullet, pointed bullet with (steel) core, was a very dangerous enemy to the tanks and the armour of these models. On 16 April another attack at the Aisne saw the loss of sixty-six tanks out of 124 (53.3 per cent). On 5 May at Laffaux, six out of twenty-two were destroyed (22 per cent), but during the battle for Flanders in 1917 things improved for the tanks. At Wytschaete on 7 June twenty-six (36 per cent) of tanks were stuck in the mud before reaching their starting point, but only two out of forty-six were destroyed by artillery (4 per cent). This worsened in the next attack on 31 July near Ypern, with seventy out of 136 tanks (50 per cent) first becoming immobilized by the mud and then knocked out by German artillery and other antitank weapons. When the Flanders battle ended in October 1917, the attacking Allies had lost 400,000 men, twice the number of the German defenders and had only gained a small terrain: 25km wide and up to 9km deep (15 × 5.6 miles). The attack on Malmaison on 23 October 1917 was started with sixty-three tanks. Of these, twenty-seven became stuck in front of their own lines and fifteen in the German lines. Only twenty-one penetrated into the depth of the hinterland. Of these, six were destroyed by the artillery.

The turning point came at Cambrai in November 1917, when 381 British tanks thrust 10km (6.2 miles) deep into the German *Siegfried-Linie*, the strong, fortified defence position of the Siegfried-Line, on a front 12km (7.5 miles) wide, while on the first day losing only fifty-nine tanks (16.6 per cent), not too many for the success gained and prisoners taken. But this was not recognized for the warning it was: the beginning of the end. The success of the following German counterthrust, planned brilliantly by Ludendorff and executed competently by the German troops, blinded the OHL to the real reason behind the defeat in the early days. They believed that capturing 9,000 men, 100 tanks, 148 guns and 716 machine guns, and winning the old positions again, made up for the loss of 10,000 men captured, together with 142 guns and 350 machine guns. Who needed tanks? The German army could fight and win without them.

*Developments of 1916 and Beyond*

But the question is: what were the Germans doing, and in particular their artillery, against the tanks all this time? This has to be analysed in three periods: from the autumn of 1916 until the turning point at Cambrai in November 1917; from this battle until July 1918; and from that point until the end of the war. Since most branches of troops were involved in this defence against the new enemy, other weapons will have to be mentioned besides the later antitank artillery.

## Autumn 1916 Until the Battle of Cambrai

The infantry had *geballte Sprengladungen*, charge packages, made out of bundles of the potato-masher *Stielhandgranate* tied together, and it also had the SmK ammunition with the hardened steel core point in the lead of the new 8mm bullets. The cartridges holding these were fired both from the German rifle *Gewehr* 98 and the heavy watercooled machine gun MG 08 and its lighter companion MG 08/15. They believed what the 119 infantry division reported: 'A troop well trained with these weapons need not fear an attack by tanks, even on a large scale.' They were wrong, of course.

The engineers had fired at armour plates on the Rheinmetall shooting range at Unterlüss in November 1916 using the low-angle mode. The light MW was able to penetrate about 10mm (0.4in) of armour at angles of impact between 60 degrees and 90 degrees with the explosive shell called *Sprengmine*. This showed promise in terms of employing the lMW for close-up antitank work, and Rheinmetall received the order to change the mount for this. They did so in a short time, designing a firing rack, where the tube and the cradle of the lMW were mounted, permitting firing in the low-angle mode up to 900m and later 1,200m (2,700–3,600ft). The engineer companies were also issued with more flamethrowers, already used at the beginning of the war against fortifications, to aim at the openings of the machine gun or observation armour. Besides this, mines and obstacles, both raised in the form of steel beams and sunken in the form of pits, were employed.

Field artillery had already shown that it could stop the tank during the attack on the Somme. Further measures included testing different shells and fuses of the field artillery to see how well they worked against tanks, and the introduction of a special antitank shell for field guns, infantry guns and short range guns, the *Kanonen-Granate* 15 *mit Panzerkopf* (K.Gr.15 m.P.), gun shell of 1915 with armour point. This was a very involved construction with the front part screwed into the body holding the explosive and the fuse in the middle. I wonder why the army personnel involved in this design seem to have overlooked the plain design of the semiarmour-piercing shells of the navy, which had been in existence since the 1890s? On the other hand, the *Wehrmacht* went into World War Two firing armour-piercing shells from the 3.7cm Pak that only had a tracer, but not a milligram of explosive. Still, the K.Gr.15 m.P. seems to have lived up to expectations, and knocked out tanks with its 170g (6oz) filling of explosive up to a distance of 3,000m (1.9 miles).

Fifty new batteries were then fielded, armed with so-called *Nahkampfgeschütze*, close-range guns, which were simply FK 96 n/A, lightened as much as possible (by dropping the lower shield and the foot bars) from the 1,020kg of the FK to 980kg (2,050–1,970lb); it was also easier to crouch down hiding in the shellcraters because of smaller but also broader wheels of 1m (40in) diameter. Another even smaller wheel on the tail was thought to help when pulling them out of shelters.

Since manufacturing of the infantry guns L/20 would take some time, it was decided to take a shortcut. At the end of 1916 the Ministry of War asked the OHL, the supreme army command, to reduce the number of FK 96 n/A being produced, and to use their parts, tubes, cradles and carriages, to mount them as makeshift infantry guns L/27. This could be done in a short time and resulted in a gun suitable as an infantry gun and for combating tanks from close range as well. Besides this, the Ministry ordered the APK to design simple small-calibre guns for the infantry, so it would no longer be helpless against tanks. These guns,

142

## Developments of 1916 and Beyond

designed along the lines of the old *Wallbüchse*, the rampart gun, a sort of oversized rifle, were to be made from seamless tubing, were easy to manufacture, and could fire their shells out of the trench over a range of a few hundred metres only. Trials discouraged the use of this gun, since the kinetic energy of the ammunition tested proved to be insufficient to penetrate the armour of tanks.

Then the APK also adapted existing small-calibre guns for the antitank role: the 2cm automatic Becker gun, the 3.7cm automatic Maxim gun, and the 3.7cm Gruson revolver gun. The 3.7cm trench guns already deployed in the front-lines and the existing batteries of infantry guns were to combat tanks together with lMW and machine guns. The drawback was that there were not enough of these. In February 1917 a French defence plan that went into detail on what to do against an attack of German tanks was captured. The OHL drew certain conclusions from this:

> Since the French believe their 37mm quick-fire gun to be the best defence against tanks and since they even believe that special ammunition for the small arms is sufficient to stop tanks, we may conclude that the armour of British and French tanks is a thin one only [no enemy tanks had yet been captured] and our means of defence against them will be effective. The most important thing is to condition our troops against the strong moral impression the tanks create.

The following battles with attacking tanks seemed to prove this assumption. The *Heeresgruppe Kronprinz Ruprecht*, army group KR, reported that three to four rounds of the *Nahkampfbatterien* had been sufficient to knock out a tank, which due to their low speed were a sitting target for the artillery, with machine guns, *leichte Minenwerfer* and *Handgranaten* mopping up the rest. This caused the OHL to believe that the present weapons for antitank warfare were sufficient and nothing else was needed; it was the wrong assumption, as it did not recognize the importance of this new weapon and by far overrated the effect of the SmK hard core bullets. Tanks captured in the meantime showed the armour to be 6–13mm (0.4–0.5in) thick. Tests ordered by the Ministry under the direction of the *Gewehrprüfungskomission* (GPK), rifle testing committee, analogue to the APK for artillery, in 1917 showed that single SmK bullets did not penetrate, but longer bursts hitting the same spot at up to 150m (450ft) did so. As the tank could hardly be expected to stop long enough for this to happen during an attack, it seemed that the machine gun MG 08 was no antitank weapon, even with SmK bullets. Thus the Ministry ordered the GPK to develop another machine gun for this role:

- with double the penetration of the 8mm SmK bullet
- with a rate of fire in full automatic of about 100 rounds in ten seconds, 600 rounds per minute
- the weapon to be carried together with its mount by two soldiers.

The calibre was thus raised to 13mm and the rate of fire increased, since this new machine gun was also to be used against enemy ground-combat aircraft. In the end this resulted in two weapons: the 13mm antitank single-shot rifle of Mauser, and the 13mm *Tank- und Fliegerabwehr*-MG, antitank- and plane machine gun. The 13mm cartridge was adapted after 1918 by the USA for their new heavy machine gun, the .50 calibre designed in 1919 by John Moses Browning, which still exists worldwide as the M2 HB (heavy barrel).

The APK also tried out another 5.7cm *Infanteriegeschütz*, really an antitank gun, built from the tube of a Belgian ditch defence fortress gun, mounted on a shortened FK 96 n/A carriage. This gun weighed 650kg (1,365lb) emplaced, fired with an increased charge up to 6,400m (19,200ft) and penetrated 15mm (0.6in) of armour plate at a range of 2,000m (6,000ft). As the recoil mechanisms, cradles and carriages coming from the FK 96 lines would have cut down the production of these, the 5.7cm gun was not introduced; their tubes were later used to arm the German tanks.

## Developments of 1916 and Beyond

The APK also experimented with the shells of both 10cm *Kanone* and 15cm sFH, showing the 10cm cannon to penetrate the armour involved from up to 4,500m (13,500ft) and the 15cm field howitzer even at maximum range.

The engineers were to be armed with special antitank mines and the infantry with *Handnebelbomben*, hand smoke bombs, and also *Nebelminen*, smoke mines, for the light and medium mine launchers.

The air force was also to participate in antitank warfare, with SmK bullets for their two aircooled Parabellum-MG, developed by Heinemann at the *Deutsche Waffen- und Munitionsfabrik* (DWM), which had the telegraph-address Parabellum, derived from their motto (after Vegetius): '*Si vis pacem, para bellum*', If you want peace, prepare yourself for war, a more elegant version of Teddy Roosevelt's 'Speak softly, but carry a damn big stick'. They also called the famous 9mm pistol, developed for them by Georg Luger from the Borchardt, Parabellum, which even today is the name of the civilian version, the military version known as *Pistole* 08 in Germany and Luger abroad. And its cartridge, the 9 × 19mm, is the most distributed hand-gun cartridge in the world, thanks to almost all submachine guns since Schmeisser's first *Maschinenpistole* 18, made in this calibre.

Back to antitank artillery. The two models of the 2cm *Flugzeugkanone*, 0.8in aircraft cannon, one designed by Erhardt and the other by Becker, were to be tested against tanks. The trench gun was neglected, no longer believed to be powerful enough, as its shells were stopped by 15mm (0.6in) armour at 150m (450ft), insufficient for the heavier tanks expected.

### From the Battle of Cambrai (November 1917) to July 1918

This decisive battle started with a new strategy, without artillery fire long ahead of the attack, to keep this a secret for as long as possible. The German long-range artillery had been blinded by artificial smoke and the British tanks were also launched out of smoke, this time against the *Siegfriedstellung*. The surprise worked and was a great success, with the tanks penetrating the fortified and prepared *Siegfriedstellung*, the Siegfried position, in a sector of 10×7km (6.6×4.4 miles). As long-range artillery could not acquire targets, the burden of antitank defence was laid solely on the shoulders of the small calibres. Field Marshal Haig reported that '… many of the hits on our tanks at Flesquières were the work of a German officer, who as last man of his battery served a field gun singlehanded, until he also was killed. This bravery caused great admiration by all ranks.' The German *Reichsarchiv*, the archives of the empire, have found out his name: It was Leutnant d.R. (of the reserve) Karl Müller, whose battery no. 6 of field artillery regiment 108 knocked out eight tanks on this day.

Cambrai brought a victory for the tank and the Allies. It also made Germany realize that the infantry urgently needed a small-calibre quick-fire gun for close-range defence.

The 2cm *Flugzeugkanone*, full automatic aircraft guns, of Ehrhardt and Becker design could now be mounted either onto the sled-type mount of the MG 08, or a special makeshift mount for antitank fire. With a newly developed armour-piercing shell they penetrated a plate of 13mm (0.5in) up to 250m (750ft). Two hundred of these guns had reached the front-lines when they were withdrawn again by the OHL in May 1918, because they believed them to be inferior to the MG with SmK ammunition. This overconfidence in the SmK bullet was a mistake, destined to be punished during the battles of summer 1918. The 2cm guns should have received a bit more attention to remove the bugs inevitable when putting aircraft weapons to ground work.

The army thought the same and continued improving the 2cm, and also started trials with a new 3.7cm gun designed by a Bavarian lieutenant colonel named Fischer. The 3.7cm Tak Fischer (Tak was for *Tankabwehrkanone*, antitank gun) also consisted of another 3.7cm tube of the Gruson-Hotchkiss revolving gun, like other makeshift guns before it. But it also had a semiautomatic breech system with a folding breechblock and sat on a captured French machine gun tripod made by Puteaux. The gun worked on the run-out principle, the tube being thrown forward by a compressed spring on firing. In the forward position the hitherto vertical hammer was levered up, hitting the horizontal firing pin and firing the cartridge. Recoil was braked by the strong spring and friction, and also by all the work it had to accomplish: open the breech, extract and eject the cartridge case. Loading a fresh cartridge and pressing the trigger again started the cycle of closing the breech during run-out of the tube and igniting the primer in forward position. The gun weighed only 78kg (167lb) and could be carried by three men, even at the double, the only way to move under enemy fire. It was drawn on a machine gun cart in more peaceful situations, broken down, but quickly assembled again within ninety seconds, the same time required for emplacement. It fired at a rate of thirty-five rounds per minute, was of simple design and good accuracy, so the APK was told to work on it. It was to start with the ammunition of the old revolving guns, which would have to be boosted for deeper penetration.

Later this gun was introduced and 2,000 ordered, of which none reached the trenches until the war's end. This was due to the stressed situation of German industry, which could not deliver before January 1919, and also the need for a newly developed cartridge. The old 3.7 Gruson cartridge of blackpowder days had been fast enough for its time with a muzzle velocity of 400m/s (1,200ft/s). But this penetrated only 13mm of armour plate at fifty paces, no longer considered to be sufficient. A new armour-piercing shell, launched at 560m/s (1,680ft/s) that crashed through 16mm (0.65in) of nickel steel at 450m (1,350ft) was the satisfying result, but was again too late.

As the 13mm TuF machine gun was considered unlikely to be ready before spring of 1918, a simple rifle for the powerful cartridge was ordered from and made by Mauser. The first of these reached the trenches in January 1918 and mass production of this *Tankgewehr*, tank rifle, began with a 30,000 order. From the end of March these *T-Gewehre* (the usual abbreviation) armed the infantry in increasing numbers. The rifle, showing 16.5kg (34.5lb) on the scales, was fired from the bipod of the MG 08/15, also to soften recoil. The bullet weighing 52.5g (1.7oz) had a steel core in its lead and departed at 785m/s (2,350ft/s) from the muzzle of the single-shot rifle. This was found to be enough to penetrate 20mm (0.85in) into an armour plate at up to 500m (1,500ft). Unhappily, the French Renault tank, with its 16mm armour, was not penetrated. The reason for this, found after the war, was simply that the 16mm penetration was made at an angle of impact of only 90 degrees, a situation unlikely to occur during a battle when the tanks were moving.

The engineers were not forgotten, receiving a new *Tankhandgranate*, antitank grenade. By spring 1918 all lMW had been furnished with a *Flachbahnlafette*, a flat trajectory carriage.

Cambrai had also demonstrated the effect of the *leichte Kraftwagenflak*, the small-calibre antiaircraft artillery mounted on motor cars. They were very mobile, fired rapidly and had a tremendous azimuth, the largest possible with their 360 degrees. They had knocked out a large number of tanks. But German industry was not able to make more of them quickly. So a number of 7.7cm FK 96 n/A *auf 4 to-Lastkraftwagen* FK on four-ton trucks, were assembled as an interim measure. These normal field guns stood on planks laid onto the loading space, traversing up to 12 degrees to each side. More azimuth was taken by the whole truck steering in this direction. The truck braked to a standstill when firing. This mounting on boards was good enough for even long-term firing, without anything getting loose.

All available Belgian 5.7cm *Kasemattkanonen*, casemate guns, were mounted in the same way. These Nordenfeldt guns were better suited for this work, since they were mounted on a pivot mount (the Belgian term for this was *affût à crinoline*, since their pivot did indeed resemble a crinoline, the wide conical frocks of the ladies of that time, reaching down to the ground). This mount gave an azimuth of a full 360 degrees. The gun was also a quick-fire gun firing cartridges, and the calibre of 57mm (2.3in) caused less recoil than the 77mm FK. The gun had an L/26.3 resulting in a tube of 1,504mm (4.5ft), which together with its Nordenfeldt folding-breech weighed 193kg (400lb). The shells left it at muzzle velocities of between 395–487m/s (1,185–1,461ft/s), depending on the charge, to ranges from 4,000–6,400m (12,000–19,200ft). The gun fired different shells, both semiarmour-piercing and armourpiercing, and canister too. The apparent failures of the enemies with their tanks made the German OHL doubt the need for this new weapon. This seemed to be proven by the success of the German spring offensive in 1918, gained without tanks. The general belief was that Germany could do without tanks and handle the enemy tanks with their existing antitank weapons. This illusion was soon to be destroyed.

## From July 1918 until the End of the War

On 18 July, the French attacked out of the woods of Villers-Cotterets with 337 tanks, mostly light ones, hidden by smoke and without warning from preparatory artillery fire. It was a tremendous success, with the German front pushed in up to 9km (5.6 miles) over a width of 50km (31 miles). The reasons for this heavy German setback were many, starting with the fact that the artillery were unable to see the targets; the infantry could not see the tanks advancing in the high cornfields; and neither the SmK bullets nor the 13mm *T-Gewehr* were effective there. The 2cm cannon had been withdrawn and the close-range batteries dissolved. The morale of the troop suffered. The German artillery, emplaced in high cornfields that blocked target acquisition and gave no cover, was surprised again as much as the infantry. It was later called *der schwarze Tag des deutschen Heeres*, the black day of the German army, despite the fact that of 300 tanks fielded on the first day, only six were in action by the fifth day. Now it really was time for the OHL to improve the antitank guns. After a conference with the Ministry of War, it ordered:

- the production of the most effective antitank weapon, the 13mm MG, should be hurried up as much as possible
- that the 2cm cannon should not be under consideration for antitank action
- that trials involving the use of the tube of the 3.7cm revolving guns in the antitank role were to be commence. The new gun should be extremely mobile in rough terrain, and must be fielded in a high number as soon as possible. At the same time, a new 3.7cm *Panzerkopfgranate*, armour piercing shell, has to be tried out.

Now the OHL really had problems. The mass production of the 13mm MG would only start in January 1919, the 13mm *T-Gewehr* was not up to the demands of the troop, and the introduction of the 2cm guns as antitank guns had been cancelled by the OHL itself. This unwise decision was now reversed again. This came too late, however, because the OHL had not realized the need for small-calibre antitank guns until the end of July 1918 and had opposed all earlier requests from the suffering troops. Had they left the 2cm Tak in the front line in January 1918, they would have had a close-range defence against tanks in the summer of 1918. After all, the 2cm shell pierced armour of 16mm at 100m (0.64in at 300ft); 14mm at 200m (0.56in at 600ft); 10mm at 600m (0.4in at 1,800ft); and 7mm at 1,000m (0.28in at 3,000ft).

## The 3.7cm *Tankabwehrkanonen*, the 1.5in Antitank Guns

Krupp and Rheinmetall were asked to submit designs of guns using both the tubes of the 3.7cm revolving guns and the carriage of the 7.7cm *leichte Minenwerfer*. The APK then tested the following guns at their firing range at Kummersdorf near Berlin: three Rheinmetall designs, one Krupp

*Developments of 1916 and Beyond*

*Rheinmetall 3.7cm* Tankabwehrkanone *(TaK), in training on a shooting range under peacetime conditions. The crew seems rather large with six men (number four from the left is an observer).*

design, the 3.7cm Fischer gun, one 2cm gun Becker II on the *Schlittenlafette* 08 of the MG 08, and one 2cm gun Ehrhardt (Rheinmetall I) on a simplified carriage of the *leichte Minenwerfer*. The 3.7cm Rheinmetall guns differed in the following ways:

1) The first had the 3.7cm tube in a reamed tube of the lMW without the trunnions. The bedding had been removed and the mounting was transported on an axle with two small wheels. The weight of the gun was 146kg (216lb).
2) In a second version of this design the gun no longer fired from wheels, but rested on a metal platform.
3) The last version weighed 175kg (368lb) and its wheels were fixed to the mounting.

The Krupp 3.7cm gun had the tube fixed to the flat trajectory lMW mount. The whole was to be fired resting on the bedding of the lMW and weighed about 300kg (630lb). This was too much because of the bedding.

All guns were tested for mobility, accuracy at 200m and 500m (600ft and 1,500ft) and armour penetration at 200m (600ft). The decision was made and the following were selected:

- The 3.7cm *Tak Krupp*, the Krupp model, with the 3.7cm tube receiving the breech of the trench gun and resting on the carriage of the lMW with large wheels. All parts not absolutely necessary had to be removed for reasons of weight.
- The 3.7cm *Tak Rheinmetall in starrer Räderlafette*, the Rheinmetall model in a fixed wheel carriage. This had the tube fixed to the carriage, without provision for recoil. Here too the breech of the trench gun was used. The walls of the mount consisted of sheet steel bolted together, which also took the axle with the wooden wheels of the lMW and held dismountable cartridge cases on both sides. It had mechanisms for elevation (–6 degrees to +9 degrees) and azimuth (21 degrees). The iron sights permitted firing up to 2,600m (7,800ft). Emplaced, the gun

147

itself (twenty-four cartridges included) weighed 175kg (370lb); the cart of the lMW on which it was transported, together with another six cases of twenty-four cartridges each, weighed 290kg (610lb). The gun now fired armour-piercing shells without a fuse, since they had no explosive inside – a solid shell, but with a tracer. Experiments to improve them by adding an explosive filling continued until the end of the war. This ammunition served to penetrate 15mm (0.6in) of armour at 500m (1,500ft); an explosive shell for close defence existed too.

Orders went out for 250 of the Krupp gun and 300 of the Rheinmetall gun, subsequently raised to 1,020. Later the Krupp version was dropped and only Rheinmetall's ordered. Of each model, twelve were supposed to be finished by September so the troops could start training with them. Up to the end of the war the troops received 600 altogether.

The Rheinmetall design was simple enough to start mass production quickly. It was transported by one horse on its lMW cart, and on the battlefield by four men pulling it along with harnesses. The process of creating a completely new gun design, manufacturing it and issuing it to the troops within two months was a unique achievement for Rheinmetall, unheard of before in all of military history. Even though it was only a makeshift weapon, it demonstrated its worth on the front. The APK evaluated this gun and agreed with the technical solution found: the very good stability in firing, the accuracy and the laying mechanisms, especially that of the large azimuth. Other experts agreed that the gun was a good solution to the problem.

- The 3.7cm *Tak Fischer* also left the Kummersdorf (translating as 'village of sorrow') demonstration with flying colours and good accuracy despite the name of the place. It could stay the way it was, French tripod and all. Only the muzzle velocity was to be raised to 550–600m/s (1,650–1,800ft/s), with Krupp and Rheinmetall both sharing their experience with Lt Col Fischer in further developing it, but 2,000 were ordered right away. Again, I wonder why all the high-ranking officers and well-paid designers did not think of the following combination: take the breech of the 5.3cm casemate guns sitting in the fortress caponniers and also sticking out of some 300 *Fahrpanzer*, and firstly stick into this one of the 3.7cm revolving gun tubes, and then in front of this fix the rifled part of a second 3.7cm tube, with a bushing holding the two together and aligned. This was often done during World War Two, when there were not enough lathes with sufficient distance between centres to work the long L/70 75mm gun tubes and the L/71 88mm tubes of the tank guns of Panther and Tiger II. In the case of Germany it would have raised the L/22 to about L/41, with a decided rise in muzzle velocity and punch against armour, even with the existing ammunition. If this new gun was then put back into the old mount, with the lower half discarded and new large wheels fixed to the top half, what a tank-smasher we would have had.

- Of the 2cm *Tankabwehrkanonen* Becker M II and Ehrhardt, which had started in life as aircraft and antiaircraft guns and were then altered for the antitank role and the Kummersdorf trials, the following numbers were ordered: 200 Becker and 700 Ehrhardt, both in *MG-Hilfslafette mit Dreifuss* 16, auxiliary machine gun mount with tripod of 1916. Of these, 200 reached the front. Though both guns were full automatic weapons, in the antitank role only single fire was permitted; the reason given being the waste of ammunition because of loss of accuracy in full automatic fire. Both weapons were fed the cartridges from metal clips holding twenty each. This ammunition was transported with the gun: 1,800 rounds in twenty boxes of ninety on one MG-*Handwagen*, handpushed cart, and on another the 2cm gun on the battlefield. Other means included transporting the gun and ammunition together on a horsecart on march, or with the gun broken down into four loads and the ammunition into another three loads for the men.

## 2cm *Tankabwehrkanone* Ehrhardt (Rheinmetall)

The double name was caused by the fact that Heinrich Ehrhardt was not only the inventor of this gun, but also held an even more important job: he was director of the *Rheinische Metallwaaren-Fabrik*, later shortened to Rheinmetall, at Düsseldorf. It had been his energy that had raised the plant from its modest beginnings to a world-famous gunmaker, rivalling the long-standing Krupp if not in size then at least in progress. His gun worked on the short recoil system, still used by semiautomatic pistols today, with the L/50 barrel moving back after firing until gas pressure had dropped to a safe level inside. Then the breech opened, the cartridge case was extracted and ejected and the hammer was recocked. In single-fire mode the breech was arrested by the trigger and stayed open, whereas in full automatic (the Ehrhardt gun could fire both ways) it went forward again, loading a fresh cartridge and firing this. This continued as long as the trigger was pulled or the twenty-round clip held cartridges.

The APK was content with the performance, reporting to the War Ministry that final inspections had shown that the guns were able to fire up to 1,250 rounds without noticeable wear or misfires. Later, the riflemaking Simson establishment, which produced the Ehrhardt based on the Rheinmetall drawings, also designed a magazine for this gun holding 59 cartridges. The sights were of the iron type and consisted of the ordinary rear sight and a circular front sight, similar to the one used against aerial targets

After the war the Ehrhardt gun was sold with the help of a Dutch company, *Hollandsche Artillerie Industrie en Handelsmaatschappij* (HAIHA), Dutch Artillery Industry and Trading Society, under the same name. Later, Steyr-Solothurn, a company resulting from the merger of Rheinmetall with the Austrian Steyr, and a Swiss manufacturer situated at Solothurn, took over both development and production, and thus the gun became famous in World War Two, both as an antiaircraft gun (2cm Flak 30), an antitank gun (2cm MK S-18-1000) and an aircraft gun (3cm MK 101 and 103).

The same was the case with the other 2cm gun, the 2cm *Tankabwehrkanone* Becker II. This had been designed on behalf of the *Waffenamt*, the weapon office, by Reinhold Becker, who ran a small engineering plant at Willich on the Rhine and had already patented a 20mm automatic gun in 1914. When the war went up into the skies, it was to be expected that sooner or later the machine guns firing the standard infantry ammunition of 8mm calibre would no longer be sufficient for aerial combat. Thus the order went to Becker, who built the gun in his own shop and had it ready in 1917. It did not even have exactly the same calibre as the Ehrhardt gun, this measuring 20mm and the Becker 20.1mm, and then fired with a twelve-round clip. On the other hand, the barrel of the Becker was shorter, with its L/40 resulting in 800mm, whereas Ehrhardt had gone to L/50 and made the barrel 1m (40in) exactly. As was to be expected, the shorter barrel gave a lighter gun, 30kg (63lb) compared to the 35kg (74lb) of the Ehrhardt. Otherwise, both fired a 140g (4.7oz) solid shot at 500m/s (1,500ft/s) muzzle velocity, out up to 2,500m (7,500ft), the Becker at a pressure of some 25 per cent higher. Sights and mounts were the same for both guns.

The Becker was of simpler construction, a simple blowback. As you need a locked breech even with a handgun cartridge like the 9mm × 19, the famous Parabellum/Luger cartridge, there obviously had to be a trick. There was: the *Vorlaufzündung*, firing out of battery. This made the gun fire the cartridge while the (really heavy) breechblock was still in forward motion. Recoil then had to overcome the forward momentum of this before it could start moving the breechblock backward. This was time enough for the gas pressure in the barrel to drop, until the projectile had left the barrel, if the latter was short enough. The APK was not as satisfied with the performance of the Becker, which, due to the great mass of its unlocked blowback breech, shook on firing.

This solution was later used by Rheinmetall when they developed the 30mm MK 108 and 55mm MK 112 aircraft guns in World War Two. The Becker also made a comeback then. Reinhold Becker had gone straight to Switzerland after World War One and

## Developments of 1916 and Beyond

joined with a Swiss manufacturer, bringing out a gun under the company name of SEMAG. When they went bankrupt, Swiss Oerlikon took over and sold the gun to the French airforce via Hispano-Suiza, which took a licence for manufacturing it, and later to the German Luftwaffe, the airforce, as MG FF.

Both 2cm guns were to increase their muzzle velocity, the Ehrhardt to 800m/s (2,400ft/s), the Becker to 700m/s (2,100ft/s). Becker also wanted to design a new mount for his gun, which would hold it steady even in full automatic fire. In any case, both guns had satisfied and fulfilled expectations, showing their earlier withdrawal by the OHL to have been a mistake.

Further measures were introduced to increase German antitank capacity before the end of the war. The 13mm MG was one of them. This does not really fall within the scope of this book, so just a brief summary will be given.

In the summer of 1918, the first patterns of the six manufacturers contacted by the GPK arrived. *Vorwerk* of twin-MG *Gast* fame, had blown this up for the 13mm cartridge (look at the Soviet 23mm aircraft gun Gsh 23, a modern copy). Rheinmetall had reduced their field and also submitted a pattern, as had *Maschinenfabrik Augsburg-Nürnberg* (MAN), which had been instrumental in developing the diesel engine and now submitted the best design of the three, which was based on the Maxim system; as a matter of fact an enlarged MG 08. No wonder both the GPK and the Ministry of War were satisfied with it, and on 13 August 1918 ordered its immediate production. The first fifty of this 13mm *Tank- und Fliegerabwehr-MG* (TuF), antitank and antiaircraft machine gun, were to be delivered in December 1918, and mass production was to start in January 1919.

Rheinmetall later came up with an improved model, but the decision had been made. The MAN-MG was to be the TuF. Even heavier tanks expected later in the war were to be combated by an 18mm (0.73in) version of the TuF. As it happened, only fifty TuF had been made by MAN when the war ended.

This was not the end of the developments in German antitank weapons in this period. The lMW was at the end of its power; both range and penetration no longer satisfied. So a light MW with greater range and a more powerful shell (mine) was designed. Since the rate of fire of this muzzle-loader was also thought to be too low now, a new 7.5cm *leichter Minenwerfer mit Doppellauf*, an lMW with double barrel, was designed. It had both a tube for firing 7.5cm (3in) shells against infantry, and a 3.7cm tube mounted on this and easily coupled to it, for combatting tanks with rounds fired at 600m/s (1,800ft/s) muzzle velocity. To save weight no breech was used, with the tube tilting to open. The cranked axle enabled this gun to fire from a low position (450mm/18in) in the low-angle mode out to 2,000m (6,000ft), and a higher position in the high-angle

*The former Belgian casemate gun was also positioned on its original high pivot casemate mounting in the front-line, to stop attacking tanks with special armour-piercing shells. Shown here is a well-camouflaged 5.7cm Nordenfeldt on a wheel carriage. When pulled over the gun the tarpaulin will hide it from any nosy enemy pilots.*

*Developments of 1916 and Beyond*

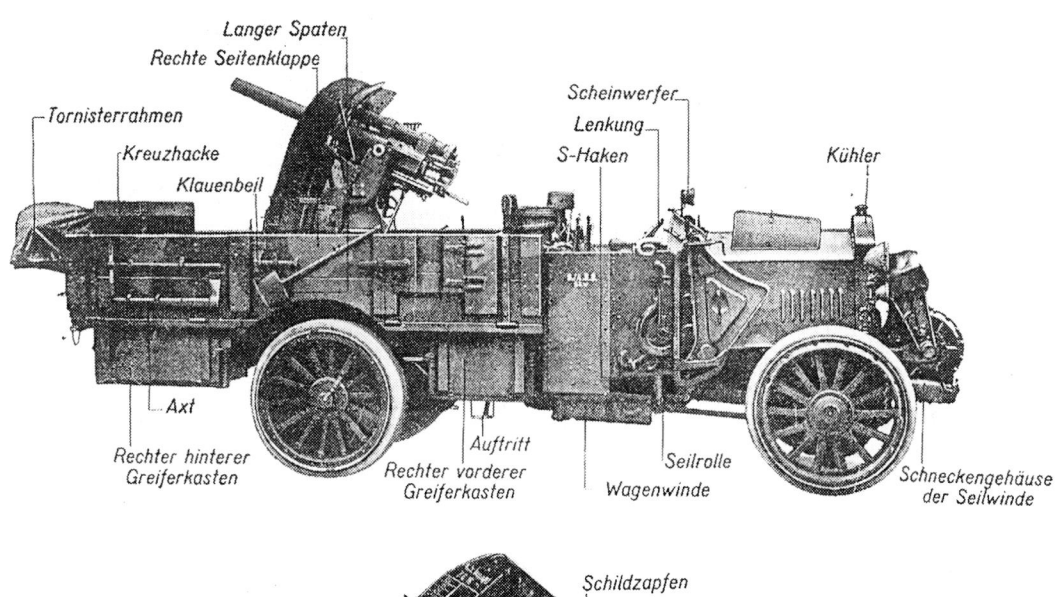

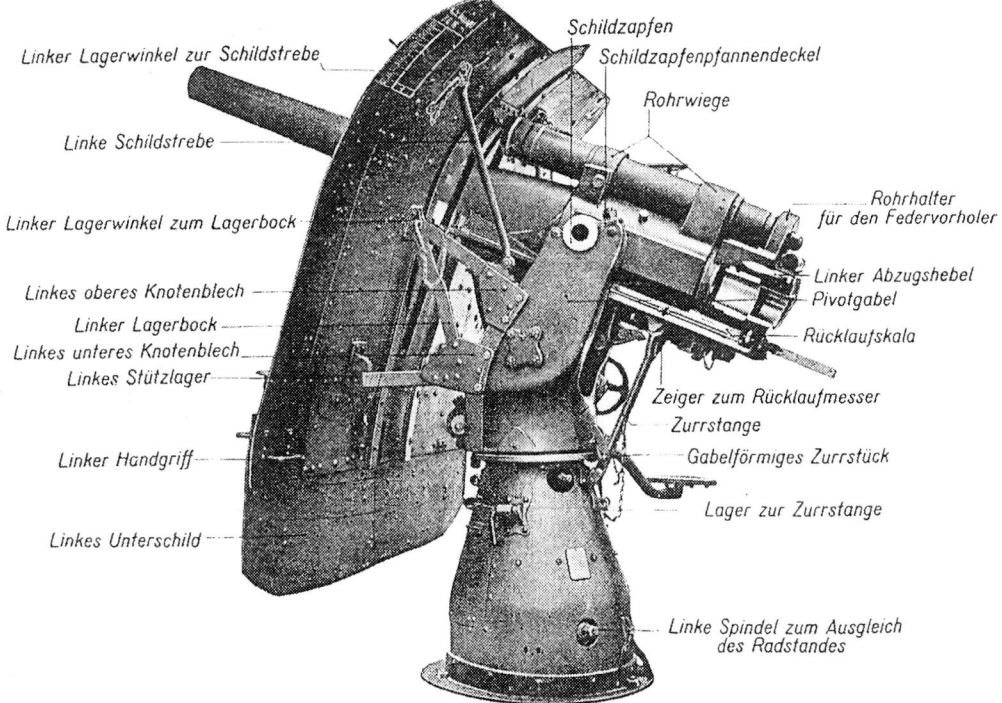

*The 7.7cm* Kraftwagengeschütz *14 by Rheinmetall was very versatile, performing both as an antiaircraft gun and later as an antitank gun. It had the largest calibre of the self-propelled antitank guns, and when firing the* Kanonengranate 15 mit Panzerkopf, *the armour-piercing shell specially designed for antitank warfare, the gunner was certain to receive the 500* Reichsmark *paid as* Beutegeld, *prize money, for destroying an enemy tank. Shown here is the 7.7cm FK 96 n/A on the 4-ton truck, together forming the* Kraftwagengeschütz 14, *the motor car gun of 1914* (top) *and the 7.7cm gun alone* (bottom).

*Developments of 1916 and Beyond*

mode up to 4,000m (12,000ft). Its weight was 350kg (735lb) emplaced because of an armour shield.

The motorized antitank guns were to be increased, too. The APK had reported to the Ministry towards the end of the war that the light motorized antiaircraft guns had been successful in warding off tank attacks. This was because of their high mobility, getting them quickly from one part of the front to another, their rapid fire and great azimuths of 360 degrees. Since the enemy would obviously be expected to enlarge his tank force, a sort of mobile antitank fire fighting had to be built up, with special guns on motorized cars, based on the model of the motorized antiaircraft gun. The following requirements seemed obvious:

- the truck should be a four-wheel drive, able to go cross-country and on country lanes at at least 12km/h (7.5mph), and on good roads at at least 25km/h (15.6mph)
- it had to be of the lowest weight possible, which excluded the use of tracks. Truck, gun, ammunition and crew had to have a weight of 6–6.5 tons at the most
- all sensitive parts of the truck had to be armoured against SmK bullets at close range
- the crew of driver plus three men had to be seated
- ammunition should amount to at least 250 rounds for the gun. This should be in the nature of 4.5–5.7cm (1.8–2.3in), with a long tube of L/40 to L/50, a muzzle velocity as high as possible (at least 600m/s, 1,800ft/s), resulting, together with a fitting weight of shell, in high penetration power. This should amount, in nickel steel, to 40mm at 800m (1.6in at 2,400ft), 39mm at 1,500m (1.2in at 4,500ft) and 18mm at 2,500m (0.7in at 7,500ft)
- the breech should be semiautomatic, the recoil long, and there should be an armour shield on the cradle and seats for layer and loader
- the gun should have an azimuth of 360 degrees, and should be able to fire ahead over the engine at least 270 degrees. High elevations were not needed; –20 degrees to +30 degrees would suffice.

According to the APK, this gun would be able to combat all present enemy tanks from a distance of 2,000–2,500m (6,000–7,500ft) and heavier future tanks with armour of around 40mm (1.6in) at 800m (2,400ft) and closer. To combat heavy tanks at long range the motorized antiaircraft guns should be armed with guns of bigger calibre than presently used. The APK also considered using more motorized transport for a quicker deployment of artillery on the battlefield, proposing motorized gun platforms for the 7.7cm FK 16 and the 10.5cm lFH 98/09.

## New Armour-Piercing Projectiles for the Light Artillery

This had been the theme of a test firing held at the Krupp range at Meppen on 4 October 1918. The results were as follows: the FK kept the K.Gr.15m.P., the armour-piercing shell, but with the explosive reduced to 100g (3.4oz) so as not to endanger their own infantry. Against heavier tanks a new solid armour-piercing shot of Krupp-design L/3.2 with tracer was to be introduced, also to be used for the infantry guns. At first, the lFH also received solid shot L/3.1 with tracer, and were going to be issued something like the K.Gr. 15m.P., with about 100g (3.4oz) explosive and a tracer. Lighter shells of 12kg (25lb) were also tested to achieve a higher muzzle velocity and flatter trajectory. The results were negative, dead ground remained almost the same and dispersion grew.

The next attempt was to improve antitank guns by different means. The first was a *verbesserte* 3.7cm TaK, an improved 3.7cm antitank gun. Meppen had shown the effect of the 3.7cm makeshift TaK to be unsatisfying against heavier tanks. In addition, the solid shot, even when penetrating the armour, had little effect inside against crew, engine, fuel tanks, ammunition and other installations. Now near the end of war, the APK demanded a new gun able to penetrate 21mm nickel steel at 1,500m (0.84in at 4,500ft). This took an armour-piercing shell of 0.6–0.8kg (1.3–1.75lb) with a muzzle velocity between 600–650m/s (1,800–1,950ft/s). The tube was to have L/40 and a semiautomatic breech and the whole gun was to weigh 130–140kg (273–310lb) without the shield, which should have a thickness of 6mm (0.25in) and be carried by one or two men. The gun itself was to be broken down into three loads for two men each, or into six loads for one man each. The carriage should be a sort of firing rack, quickly mobilized and pulled by two men.

Krupp, Rheinmetall and the engine-maker Henschel at Kassel were to submit designs. Rheinmetall did so on 29 October 1918. This was for a TaK of higher power. The 3.7cm gun L/45 fired a 750g (1.5lb) shell at 680m/s. It had a semiautomatic breech and its weight, emplaced, was 250kg (525lb), including twenty cartridges. It could be broken down into four loads for carrying. This gun was still unfinished at the end of the war.

## The 5cm TaK

The APK had realized that the time of 3.7cm antitank guns would be over when tanks shielded by 50mm (2in) armour plates attacked. A larger calibre of about 5cm (2in) seemed to be required. Krupp had already designed such a gun to arm the planned *kleine Kampfwagen*, small tank, which was a 5.2cm (2.1in) L/40 gun, firing a shell of 1.75kg (3.7lb) at 600m/s (1,800ft/s).

In September 1918 the OHL asked for a special shell for both the 10cm *Kanone* and the 15cm sFH, so that they could combat any tanks which had broken through the front lines. The idea was for a shell:

- which would not endanger their own troops through its splinters
- with a very flat trajectory for a better chance of a hit, if necessary achieved by lightening the shell
- which could be well observed with the aid of a tracer.

Since development was known to take some time, at first 10cm and 15cm shells were emptied of their explosive and their fuses. Later, the following antitank shells were intended: a 10cm shell with tracer, 100cc Bromaceton (tear gas) and the Gr.Z. 14 n/A (a new model of the shell fuse of 1914); and the 15cm shell Gr. 17, 30kg (63lb) in weight, with tracer and also 100cc Bromaceton and Gr.Z. 14 n/A.

Thus ended the fight against the ultimate enemy of the German army: the tanks of the Allies. The Reichswehr was not permitted any tanks and so had little chance to follow up the developments and plan remedies for them. The first antitank gun fielded in Germany after World War One was again of 3.7cm calibre, at a time when, for example, the guns in the caponniers of modern Belgian forts such as Eben Emael were already of an impressive 60mm (2.4in) calibre; a calibre not even reached by the second German *Panzerabwehrkanone* (PaK), as the former TaK was now called. And when during the war the inevitable leapfrogging started between armour and antitank gun, it reached an unbelievable 88mm, and even 128mm (5in); guns immobilized by their own weight.

Today, the modern tank is almost immune to warheads of even the shaped charge type, thanks to its multilayer mixed armour of Chobham type, and also additional 'reactive armour'. This consists of brick-shaped boxes fixed all over the tank, containing a sensitive fuse and a thin foil of explosive which, when detonated, throws a steel plate across the 'jet', the copper particles of the cone, speeded by the warhead up to 10km/s (6.6 miles/s), or twelve times the muzzle velocity of a modern tank gun at around 1,800m/s (1.1 miles/s). But the jet is disturbed and no longer powerful enough to penetrate a tank. Only the kinetic energy of a really big gun – at least 120mm – will do the trick, with a lot of megajoules of power transferred from the propellant onto the 'flying nail' of the long rod penetrator: 2ft of fin-stabilized *Staballoy*, 1in in diameter. And to visualize the power of such tank guns, which will penetrate 1m (40in) of rolled homogeneous armour (RHA) like butter, just drive the tank carrying it onto a hill so that elevation amounts to some 50+ degrees, and fire. You will find out that you are driving around with a Paris gun of your own, with the penetrator impacting some 120+ km away. This is modern ammunition.

## Antitank Guns

| Gun model | Calibre (in) | Weight Empl. (lb) | Tube Length (in) | Shell Weight (lb) | Muzzle Velocity (ft/sec) | Max. Range (ft) | Elevation/ Azimuth (degr.) | Remarks |
|---|---|---|---|---|---|---|---|---|
| 2cm Tak Becker M II | 0.8 | 63 | 32.2 | 5oz | 1,500 | 7,500 | +45/360 | 120 rpm (1) |
| 2cm Tak Ehrhardt | 0.8 | 73.5 | 40 | 5oz | 1,500 | 7,500 | +45/360 | 120 rpm (2) |
| 3.7cm Tak Krupp | 1.5 | n/a | 31.8 | 1 | 1,518 | 7,800 | n/a | (3) Krupp |
| 3.7cm Tak Rheinm | 1.5 | 368 | 31.8 | 1 | 1,518 | 7,800 | +9/21 | (4) Rheinm |
| 3.7cm Tak Fischer | 1.5 | 164 | 31.8 | 1 | 1,518 | 7,800 | n/a | (5) Fischer |
| 5.7cm KasK Kw | 2.2 | 193 | 60.2 | 5.7 | n/a | 19,200 | +21/360 | (6) Nordenfeldt |
| 7.5cm IMW Doplauf | 1.5/3.1 | 735 | 16/31.8 | 9.7/1 | 885/1,200 | 600/1,800 | +75/10 | (7) Rhein/ n/a |
| 7.7cm NahKampfG | 3 | 1,869 | 83.2 | 14.4 | 1,395 | 23,400 | +15/8 | (8) Krupp |

*Remarks:* (1) improved version of the aircraft model, only to fire single rounds, weight of gun only; (2) Rheinmetall-made, also only to fire single rounds, weight of gun only; (3) on carriage of the IMW; (4) special carriage; (5) not fielded; (6) ex-Belgian fortress gun mounted on 4-ton trucks; (7) *Doppellauf* = twin tube, a new 75mm! light *Minenwerfer* combined with a 37mm antitank gun on top. Figures are for MW/ATG; (8) *Nahkampfgeschütz* = close combat gun, FK 96 n/A on 4-ton trucks.

n/a = no data available.

## TANK GUNS

We have learned that after the first encounters with British tanks, especially at Cambrai, which, through an energetic counter-attack was turned from a German defeat into a draw, the German OHL and (almost) all of the German army, was of the firm believe that: who needs tanks? They had concentrated on antitank measures, but not on developing tanks of their own design. Thus at the end of the war, which Germany lost to the tanks of the Allies, almost no German tanks had appeared on the battlefield. Then, finally, there were plans, big ones. But 'paper is patient', as an old saying has it.

Yet it need not have happened this way. We shall look deeper into this matter, since contrary to Britain's thousands of tanks, Germany only fielded twenty of its own design and manufacture, of which ten were lost in action and the rest cut up for scrap after the war. Only one survived, far away in Australia.

Before looking at the German tank guns, we have to start with the German tanks. In the 1890s a German army officer, Lt Col Layriz, wrote a book called (translated) *Mechanical Traction in War*. Its emphasis was on steam traction, actually used before in wars, starting with British Fowler road engines pulling Prussian guns in the war of 1870–71, and ending with the armoured road trains ordered by Lord Roberts for the war in South Africa. Of these, Layriz reported on the state of the art, and wrote:

> The armour consists of steel plates a quarter of an inch thick, specially hardened by the Cammell process. The plates will resist direct rifle fire at twenty yards, and are impervious to shrapnel or splinters of shell. Needless to say, however, they will not stand direct shell fire. Exactly to what

purpose the train will be put is still largely a matter of conjecture, but it is interesting to know that a mechanical problem of some difficulty has been solved, and that Lord Roberts will soon have under his command six well-armoured trains capable of following any ordinary road, or of striking boldly across the veldt.

He wrote about existing armoured transport and did not yet see the approach of something like a tank, defined as an armed and armoured motorized fighting vehicle, travelling on tracks cross country. This was done by other men.

One of them was the German General Richter, who in 1907 suggested that guns with armour shields, mounted on motor cars, should accompany the attacking infantry.

Finally, in 1911 an Austrian First Lieutenant named Burstyn proposed an armoured motorgun, running on tracks, which he named *Gleitband*, sliding band. On 28 February 1912 he received the patent number 252 815 for this in Germany, but one year later the KM, the ministries of war both in Austria and Germany, refused to accept his invention.

Motor cars had been tested in the combat role during the German manoeuvres of 1910 and found unsatisfying, leaving the motor car solely in the transport role. Thus it took the encounter with British and Belgian wheeled armoured scout cars in the autumn of 1914 to move the OHL into action. On 22 October 1914 they demanded the development of special German wheeled armoured cars with four-wheel drive, not constructed as armoured boxes placed on the chassis of passenger cars. The three resulting patterns offered by the competing companies of Ehrhardt, Daimler and Büssing were armed with three machine guns each, but no cannon, and were sent to Roumania to gather experiences in combat. In 1916 they returned to the west front, now as *Panzerkraftwagen-MG-Truppe*, armoured car machine gun troops, and in 1917 another twelve Ehrhardt (Rheinmetall) cars joined them. Later, Italian, British and Russian armoured cars were also integrated, and all satisfied in this role until the end of the war.

Britain had started work on tanks based on the Caterpillar track system and the Burstyn patents in 1915, with Colonel Sinton and W. Tritton in charge. The first appearance of British tanks in 1916 at the Somme, and the little success gained by them, had caused the German army to believe, until Cambrai in 1917, that Germany needed no tanks at all, since the army could fight and win without them. But the supreme army command of the OHL took the battle of the Somme seriously enough to report, on 11 October 1916, on the strength of the tanks to the KM, Ministry of War, and to demand the development of German tanks, to be mass-produced once a suitable design had been found. For this task the OHL also wanted the Austrian industry to join in. After a conference in Berlin, the OHL was informed by the KM on 11 November 1916 that their own cross-country vehicles had been tested, but had not proved satisfying enough to armour them. Therefore these first twenty vehicles were used only as trucks. (These were the results of a development by the *Verkehrstechnische Prüfungskomission* (VPK), a committee on the level of the well-known APK and GPK, and doing the same sort of work as these, but in the transport sector. They wanted a cross-country vehicle, at first for supply and later also for combat. Results with the so-called *Bremerwagen*, Bremer car, moving on four tracks, had not shown satisfying results so far.)

On 13 November 1916, the KM then gave orders to the APK, GPK and VPK to start work, together with the German motor vehicle industry, on the first German tank: the A7V, a code name derived from the office of KM responsible for this – *Abteilung 7, Verkehrswesen*, department seven, transportation. (Later, the cat was let out of the bag and the tank was called *Sturmpanzerwagen* A7V, armoured assault car.) A technical committee was formed, the head of which was a captain of the reserve, *Oberingenieur* Joseph Vollmer, a man who was well qualified for the job through his earlier civilian career. Vollmer was the founder of the *Automobilgesellschaft*, the Automotive Society, in Berlin 1907, and was put in charge of around forty design engineers of the automobile

155

## Developments of 1916 and Beyond

industry. They finished this design within an incredible four weeks, which seems to have been insufficient time to consider not only the technical aspects, but also the tactical aspects of such a vehicle. The A7V was cursed with poor cross-country manoeuvrability, due to its short tracks ending a long way behind the front end, and at 30 tons it was too heavy from the start. But it was the only design Germany had, and not having time enough to continue looking into the matter, the KM ordered 100 of these right away on 29 December. Shortly before this the OHL had demonstrated how serious they took the business of motors, cars, trucks, and everything of a similar nature, by creating a new office for it: the *Chef des Feldkraftfahrwesens*, Chef Kraft; in brief, the Chief of Field Motorization. A wooden model was built of the A7V and demonstrated on 6 January, and 100 more A7V chassis were ordered, mostly for other designs, such as ammunition transporters.

Then things slowed down. In February the OHL asked Vollmer to make the A7V proof against field gun fire, which he did. But Chef Kraft now changed his mind and asked for a track of the British-type, running over the top of the tank and resulting in much higher climbing ability than the A7V had, but also making it bow-heavy. This was to be called the A7V-U, 'U' for *umlaufende Kette*, this form of track design. Captain Wegener, on the staff of Chef Kraft, began to design a super-heavy tank, the *K-Wagen*, the K-car. In the meantime, the KM asked for another 100 A7V on 31 March, and on the same day Chef Kraft ordered the design of the *K-Wagen*. And while the chassis of the A7V was demonstrated to the KM at Berlin on 4 April, the OHL revoked their own vote giving top priority to the A7V. Then, on 28 June 1917, Chef Kraft asked for ten *K-Wagen* to be built, revoking orders for the building of another ten A7V. Small wonder that it was October 1917 before the first A7V was finished and demonstrated to the KM.

As a result of their experiences at Cambrai, on 23 November the OHL (which seems to have thought more realistically than the optimistic troops) ordered all German tanks to be armed with cannon, machine guns alone no longer being sufficient, and finally put the A7V in top priority on 12 December.

On 2 February 1918 came the final decision. The A7V was demonstrated – together with other vehicle models – to the Kaiser. Chef Kraft called it '... not exactly useful in the field', and this led to only twenty of them being ordered in the end. Then news came of a new French light tank having been observed on 16 April 1918. This resulted in contact with Krupp about the construction of a similar

*The only successful German tank fielded was the A7V. This was armed with a former Belgian casemate gun of 5.7cm calibre, mounted in the bow.*

## Developments of 1916 and Beyond

*The next design by the same Captain Vollmer was to have less success. It was a tank styled after the British Mark IV and named the A7V-U, for the* Umlaufkette, *the high tracks, which sported two guns in sponsons.*

light tank, for which Krupp came up with a design on 22 May. Now the KM had to confess that they had developed one such of their own, the SmK-proof, L.K. II, *Leichter Kampfwagen II*, light battle tank II. This was demonstrated on 13 May and led to an order of 580 LK II on 26 May, and also for a consolation twenty of the Krupp design.

In the meantime, Chef Kraft had had another idea and ordered the design of another heavy tank named *Oberschlesien*, upper Silesia, now Polish, to be built by the *Oberschlesien-Eisenindustrie*, the O.S. Iron Industry, at Gleiwitz. Now the plan was to manufacture all four different tanks, and more mix-ups were to follow during the rest of the war, until it was finished by the 1,000 British and 2,000 French tanks.

Back to the A7V and its guns. At first the A7V was to be armed with MG and 2cm guns of Becker design. But since the effect of a single 2cm shell was deemed insufficient against the typical tank targets such as machine-gun positions, tanks and so on, and the 2cm guns had also not yet reached perfection, the transport people suggested arming the A7V with 7.7cm FK 96 n/A. The APK refused, since they felt that the combined weight of this gun together with its ammunition would be too much for the tank, already weighed down by its heavy armour. In addition, the system of the FK 96 n/A had the gun tube recoil for 750mm (30in), and space for this was not found in the fighting compartment. Also, the cradle would protrude from the armour, which would make it very vulnerable, even under small-arms fire. It would have meant a complete redesign of both cradle and recoil mechanism, and this was judged impossible if the tank was to be fielded in the time frame desired by all.

The APK then asked for the installation of the *Belgische* 5.7cm *Kasemattkanone*, the Belgian 2.3in casemate gun. This only needed 150mm (6in) for recoil, which allowed cradle and recoil mechanism to be placed within the protective armour. This gun already served in the mobile anti-tank role, having been mounted on 4-ton trucks and furnished with special armour-piercing shells. Thus it had already proven its ability to combat enemy tanks successfully. There were changes necessary, however. The original Belgian fortress mount of the *crinoline*-type had to be exchanged for another design, since this gave little elevation, and a tank gun would have to fire even if the tank was inclined to one side or slanting downhill. A new simple makeshift mount was quickly designed by the bureau of artillery designs, with open iron sights for the first A7V. This was later exchanged for the final design, a pivot mount with telescopic sight. This was actually one of the field guns laid on its side.

At first, the supply of ammunition was 180 cartridges, containing 100 shells with *Aufschlagzünder mit Verzögerung* (Az mV), delayed impact fuse, forty shells with *Panzerkopf*, armour points, and forty cartridges with canister. The shells with the

157

## Developments of 1916 and Beyond

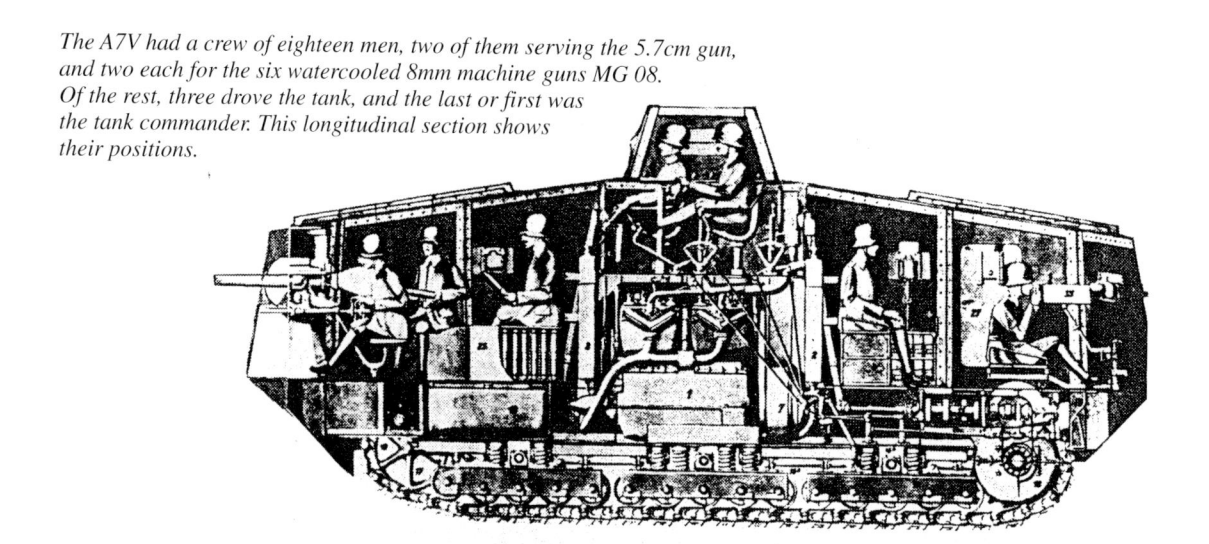

*The A7V had a crew of eighteen men, two of them serving the 5.7cm gun, and two each for the six watercooled 8mm machine guns MG 08. Of the rest, three drove the tank, and the last or first was the tank commander. This longitudinal section shows their positions.*

delayed fuse were destined for use against troops in the open or behind cover, for example a wall. The fuse replaced the normal time fuse, which could not be set in a moving tank. So the shell was fired at a shorter range, in front of the target, which made the fuse start working, but with the inbuilt time delay still preventing explosion. The shell then bounced until over the target where it detonated. The OHL objected to this fuse, since they held that for the shell to bounce, the angle of impact had to be favourable, and they wanted the Az mV fuse exchanged for a simple impact fuse. (The philosophy of range safety in NATO today holds that an angle of impact of 30 degrees or less *may* cause ricochets, but by no means guarantees them.)

The APK at once went on to show that they were right, holding test firings in a swampy terrain filled with shellholes. Over ranges of around 2,500m (7,500ft) it was shown that of forty rounds fired, thirty-six ricocheted properly. Since tanks were not expected to fight at longer ranges, the Az mV fuse seemed the best choice. But there were not enough of them. So the KM ordered the tanks to be filled with 55 per cent of shells with the Az mV fuse, 25 per cent of shells with super-quick fuses (without delay), 10 per cent armour-piercing shells and 10 per cent canister. The base fuse of the semi-armour-piercing shells, for detonating the explosive inside, was later slowed down, with the delay increased from 0.25 to 0.5 seconds. Now the shell could penetrate thicker armour of 20mm (0.8in) at 2,000m (6,000ft) and still detonate fully inside the tank, not half outside, with better effect.

The A7V was fielded for the first time in the spring offensive at St Quentin on 3 March 1918. Of the ten German tanks only half were A7V, the other five captured British tanks. Later, the A7V were judged to have been the better half. The Chef Kraft, by now well known to us, wrote the following in his report on the A7V of 5 May 1918.

1. Arming the tank with guns has shown to be very useful when suppressing enemy strongholds and fighting enemy tanks.
2. Rate of fire is sufficient, even at strong enemy counteractions.
3. Ammunition is of good effect and sufficient in number. The canister showed the best effects.
4. The periscope for aiming is a small one only, the field of vision wanders off during movements of the tank. This makes the acquisition of targets very difficult; when the visibility is impaired even impossible. All departments have asked for a return to the simple open iron sights.

## Developments of 1916 and Beyond

5. The observation openings are positioned too deep on some of the tanks, so that the gun commander can only observe in the sitting position, thus hindering the loading gunner.

The last remark led to an open sight of the shotgun-type being added, which was well suited due to the similarity between canister firing and shooting a shotgun.

On the A7VU *Wagen*, the A7V car with *Umlaufkette*, with tracks running forward over its top in the British style, two of the 5.7 Belgian casemate guns were mounted. Since the bow could not be used for one mounted in the normal position because of the tracks protruding too far for a reasonable azimuth, the guns had to be mounted in sponsons on each side, another similarity to the British system. Only one tank of this pattern was built. The other data were similar, but not identical to the normal A7V, with a length of 7.35m compared to 8.4m for the -U (22–24.2ft); a width of 3.06m compared to 4.7m for the -U (9.2–14.1ft); a weight of 30 tons against 40 for the -U; a range of 35km (22 miles) or six hours for both; two Daimler four-cylinder engines at 100hp each, giving a speed of up to 15km/h on the road for the A7V, and up to 10 for the -U (9.4/6.3mph); a crew of eighteen men each; armour of 15mm on top, 29mm at the sides and 30mm in front (0.6in; 0.8in; 1.2in) for both; armament was 1 × 5.7cm plus six machine guns for the A7V and 2 × 5.7cm plus four machine guns for the -U.

### The *Grosskampfwagen* or *K-Wagen*, the Big Combat Car

The *Grosskampfwagen* was a giant compared to the A7V, which itself dwarfed the light French tanks. Compared to the 30, and later 40 tons of the A7V, the *K-Wagen* was 150 tons, five times as much, moved by two Daimler aircraft engines of 600hp each. For railway transport it had to be broken down. It was 13m (39ft) long; 3.1m (9.3ft) wide, with 6m (18ft) at the sponsons and 2.85m (8.5ft) high. Its 150 tons, of which a good deal was the result of double armour plates of 20mm each, a total of 40mm (1.6in) in front, 30mm (1.2in) at the sides, and 20mm (0.8in) on top. This monster reached 'speeds' of up to 7.5km/h (4.7mph).

Its armament was impressive: no less than four 7.7cm guns mounted in pairs in sponsons, and two machine guns were handled by the crew of twenty-two men. These guns were not the normal FK 96 n/A, because this would again have required too much space inside for its recoil. Thus a recent adaptation of these for the caponniers of the *Festen*, the modern fortresses after 1894, of the same calibre were used, which recoiled only 400mm (16in). For each gun, 200 rounds were stored aboard; 800 altogether.

Other guns had also been investigated: the 10.5cm lFH was out because its shells lacked in antitank capacity; the 8.8cm cannon of antiaircraft and the navy could not be withdrawn from its positions; and the 7.7cm FK 16 was dropped because it did not fire quickly with cartridges, but slower because of its separate loading ammunition. Two of these super-heavies were finished at the end of the war and then scrapped, like all material of the German army.

In 1917, the OHL also ordered the repair of the British tanks captured in high numbers to increase the tanks available on the front. The German/Belgian 5.7cm ammunition was of the same calibre as that of the two-pounder British guns, but with a cartridge case about 100mm (4in) shorter. This problem was solved quickly by the installation of Belgian and Russian 5.7cm guns formerly deployed in fortifications, of which a lot were available, in these British tanks, and these were furnished with 200 rounds per gun.

### The *Leichte Kampfwagen*, the Light Tank

The *leichte Kampfwagen* had been recognized by the Germans in 1918 to be the better solution, since the early big heavies of the enemy tended to be knocked out by German artillery sooner, being both a slower and a larger target than the later smaller and faster French version. The VaKraft responsible

*Developments of 1916 and Beyond*

for motor vehicles had designed a similar light tank. But the APK insisted that arming this only with machine guns was not sufficient. A gun was needed to combat enemy tanks. After the disaster in the middle of July 1918, when the French penetrated the German lines at Villers-Cottererêts using their newly designed small Renault tanks, the OHL demanded the immediate mass production of such tanks. A makeshift design was tested and 900 of these ordered, to be manufactured using parts of passenger cars. They would have been ready in the spring of 1919. Of the three available guns – 2cm, 3.7cm and 5.7cm – the 2cm and the 3.7cm were declined because they were felt to lack punch, and the 5.7cm gun, already well proven by this time, was selected. This was to be mounted in only 40 per cent of the tanks, the other 60 per cent being armed with machine guns.

## The 5.2cm *Tankkanone* L/40 Krupp and the *Kleine Sturmwagen*, the Small Assault Car of Krupp

Krupp had accepted the arguments of the APK regarding the 3.7cm gun and the light tank, and designed a 5.2cm (2.08in) tank gun. They offered this to the APK on 15 August 1918, explaining their reasons for choosing this calibre: it was more effective against tanks because of its heavier shell, as well as against positions because of its more explosive filling than a 3.7cm. With an L/40, the tube was 2,080m (83.2in) and weighed 160kg (336lb), with the semiautomatic breech another 12kg (25.2lb). It fired a shell of 1.75kg (3.7lb) at a velocity of 600m/s (1,800ft/s). This gun was mounted in a new way: an armour cupola, able to take full azimut of 360 degrees. In this cupola the gun rested in a sort of fork mounting, in which it was moved up and down by the elevating mechanism. The whole cupola including the gun weighed exactly 700kg (1,470lb).

Together with this 5.2cm gun, at the demand of the OHL Krupp submitted a new tank design for this gun: the *kleine Sturmwagen*. With its crew of four and a weight of 7.7 tons it could travel up to 12km/h (7.5mph). It had hooks front and rear for towing infantry guns. The crew was transported on small platforms fixed to the sides of the tank. Recoil of the 5.2cm gun was too long (55cm/22in), necessitating a large armour cupola of great weight. Krupp was reasonable – after all, they had no experience in the tank design business – and promised to exchange their cupola for a design preferred by the APK. Then the OHL got involved: first they wanted the Krupp-*Sturmwagen* rearmed with a 13mm MG only, then, after Krupp had redesigned the tank accordingly, the OHL insisted on a 3.7cm gun. The APK, still holding fast to the 5.2cm gun, lost this battle. All they could get from the OHL was their agreement to a higher muzzle velocity for the 3.7cm gun for better penetration of armour plates. This resulted in the 3.7cm *Tankgeschütz Krupp*, the 1.5in tank gun of Krupp.

## 3.7cm *Tankgeschütz* Krupp

This gun was to receive a semiautomatic falling breechblock and low trunnions, possible because the recoil cylinders were no longer arranged in the traditional way underneath the gun, but on both sides of it. The tube weighed only 47kg (95lb), recoiled for 250mm (10in), and fired the 3.7cm projectile of 600g (1.3lb), later 700g (1.5lb) solid shot with tracer at a muzzle velocity of 600m/s (1,800ft/s) at first, and later 700m/s (2,100ft/s). The range was 4,300m (12,900ft) at an elevation of 15 degrees, with the complete elevation ranging from –20 degrees up to +30 degrees. The gun also fired canister and was later to receive shells, too. Both this gun and the Krupp-*Sturmwagen*, of which eighty-five had been ordered for the spring of 1919, were not finished by the end of the war.

Of all German tank models, only a single A7V survived, and this is in an Australian museum. So in 1987 a committee was founded at the German Army Office at Cologne, consisting of the military and the industry, aiming to do something for posterity, something sadly neglected in the *Bundeswehr*, the German Forces, ever since their formation in 1956 for NATO. The aim was to build

## Tank Guns

| Gun model | Calibre (in) | Weight Empl. (lb) | Tube Length (in) | Shell Weight (lb) | Muzzle Velocity (ft/sec) | Max. Range (ft) | Elevation/ Azimuth (degr.) | Remarks |
|---|---|---|---|---|---|---|---|---|
| 5.7cm PzW-G | 2.2 | 405 | 60 | 5.67 | 1,461 | 19,200 | +20/45 | (1) Nordenfeldt |
| 7.7cm PzW-G | 3 | 1,869 | 83.2 | 14.4 | 1,395 | 23,400 | +20/45 | (2) Krupp |
| 3.7cm TankG Kr | 1.5 | 98.7 | n/a | 14.7 | 2,100 | 12,900 | +30/15 | (3) Krupp |
| 5.2cm TankK L/40 | 2 | 596 | 83.2 | 3.68 | 1,800 | n/a | +20/360 | (3) Krupp |

*Remarks:* Also under consideration: the 2cm Becker MK. Further designs by Rheinmetall and Krupp not finished by the end of the war. All tank guns fired fixed cartridges. (1) ex-Belgian fortress gun firing two different charges; (2) a special adaptation of the FK 96 n/A for casemates with shorter recoil; (3) semiautomatic.

n/a = data not available.

a replica of the A7V, to show the importance of the tanks in modern warfare. The tank has now been built and may be admired in the *Panzermuseum*, the Tank Museum, situated in the small village of Munster, home of the German *Panzertruppe*, the tank force, since the days of General Guderian.

## AIRCRAFT GUNS

We have already encountered both examples of German aircraft guns firing ammunition containing explosives: the Becker gun and the Ehrhardt gun. (As a matter of fact, the Becker initially fired only solid shot, but in Germany the machine gun ended below 20mm, where the *Maschinenkanone*, the machine cannon, the automatic gun, began.) They both served as quick-fire weapons in antiaircraft and antitank roles first, and in the end they were even considered as tank guns – quite a career.

Let us take a quick look at their origins. The war in the air had begun on a friendly, polite level, with the pilots saluting each other, even those on opposing sides, wanting to damage only those on the ground by dropping handfuls of *Fliegerpfeile*, pilot arrows (dartlike pointed steel missiles) onto them. Then someone took his firearm up into the clouds and the shooting started, one side wielding the .45 semiautomatic Colt hand gun, the other an 8mm selfloading rifle, named after its Mexican constructor, General Mondragon. More firepower was gained by dragging machine guns into the air, though these caused problems when fired – inevitably – through the arc in which the airscrew rotated, managing to hit this as well and thus downing their own aircraft. This was solved on the German side by the genius of the Dutch aircraft designer Anthony Fokker, who analysed the pattern found on a French plane shot down with rifles (ironically, it was the plane that the inventor Garros had flown himself) and then within forty-eight hours designed a solution that was later found on most planes, even at the beginning of World War Two.

The machine guns were already cooled by the air rushing along them, and their rate of fire kept increasing, as well as the number mounted on the plane: one soon became two, with Immelmann asking for and getting a third. This density of bullets was also beefed up a good deal by the unique design of the German 8mm Gast-MG, or rather two of them, mounted in a single housing, with the recoil of one breechblock speeding up the closing of the other one connected to it. (Later, the same was tried by Gast in 13mm, and today the ex-Soviet 23mm Gsh-23 twin gun is mounted on most of the CIS-planes.) Tracers – a British invention of

*Developments of 1916 and Beyond*

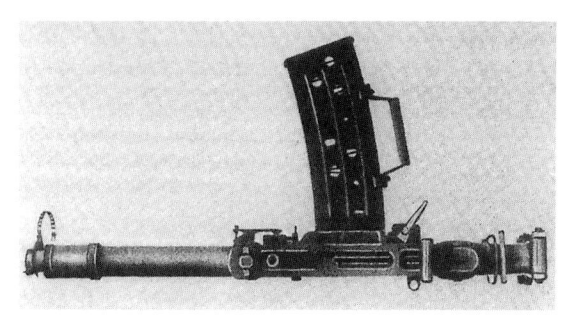

*2cm Becker-*Flugzeugkanone, *aircraft gun:* (above) *fixed mounting with twelve-round magazine inserted;* (right) *in pivot mounting for flexible firing.*

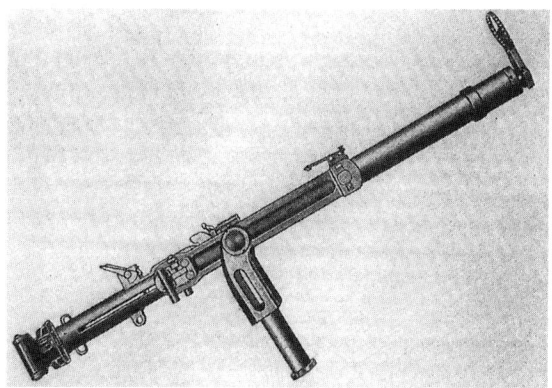

before the war, but only introduced in 1916 as 'sparklets' – helped to hit the plane, and steel cores partly replacing the lead of the bullet helped to penetrate to the vital parts.

The next step was to provide the planes with armour, a typical example being the German *Gotha* bomber, on which the vital parts of engines and pilots' seats were covered with armour plate. This took to the air from 1917 onwards and coordinated the efforts of the German army and airforce towards developing a machine gun of bigger calibre, for combatting both tanks and planes. It was named *Tank- und Fliegerabwehr-MG* (TuF), the antitank and antiaircraft machine gun, had a 13mm (0.52in) calibre and its design was based on the Maxim by MAN of diesel engine fame. Its 250 different parts were manufactured by no less than sixty different producers and assembled at MAN at Augsburg.

The next logical step was to take automatic guns up on planes, firing explosive shells. That was when Becker and Ehrhardt got their chance.

The small-calibre automatic gun had been in the air before, not on a plane, but on a dirigible. Already in 1910, a German, Captain de le Roi (of Huguenotic lineage), while giving a lecture to the *General der Verkehrstruppen*, General of Traffic Troops, Freiherr von Lyncker, about the further development of aircraft, suggested that planes should reconnoitre and fight dirigibles. Captain de le Roi had been in charge of a new department for aeronautics, founded by the KM, the Ministry of War, in 1907, and was the top authority on this matter. Incidentally, 1907 was the year in which the Wright brothers offered their planes to the KM for a second time, having already done so in 1906. But the words of de le Roi were commented on by one Major Gross, chief of the *Luftschiff-Abteilung*, the dirigible-section, who told the General: 'Quite nice, what this young captain said, but planes will never be of any importance in war.' De le Roi also suggested dropping acid or small bombs on to enemy dirigibles or blimps. General Lyncker had already feared such a fate for his dirigibles, and in 1909 called in a commission which was to propose means by which to turn the dirigibles into combat airships. This meant firing shells against operative targets such as depots, forts, bivouacs and staffs, and also required hand guns and machine guns for self defence. At this time, however, not all of the old staff officers were convinced by the machine gun, which had been introduced because of pressure from Kaiser Wilhelm II. In 1899 he had paid for the first MGs to arm his *Garde*, the Imperial Guard troops, out of his own pocket. Two years later, MG 01 was introduced, having to be continually improved, over MG 03, until the final model was fielded in 1908 as MG 08. The GPK, the rifle-testing commission, was ordered to take up the matter of using machine guns on planes.

Bombs were also a new development. There had been the clever design of the two Uchatius brothers in 1849, both then first lieutenants in the

Austrian army. They had fixed a 22.2kg (46lb) bomb of their own construction to a small montgolfier, a balloon which rose by heating the air inside. This drifted all alone – with no crew – over besieged Venice, until the heat of the charcoal burned for fuel ignited a blackpowder charge, cutting the rope which held the bomb to the miniballoon. The cast iron bomb dropped onto the target, the arsenal of Venice, and exploded when the piece of fuse had finished burning through.

A new development was then undertaken as a private venture by the explosive manufacturer of the *Carbonit-Sprengstoffwerke* in Schlehbusch, who had received the impetus from another young lieutenant, Mackenthun, in 1909. The same Lieutenant Mackenthun was also the first pilot to demonstrate the dropping of a bomb of 3.5kg (7.7lb) in 1911 at Doeberitz; the display impressive enough for the first orders for bombs to be written the next day.

It was not taken to be a sign of a low mind to think about defending your own country in those years. For example, Michelin (of the tyre fame) in France founded the *Prix-Michelin* in 1911. This reward was for the pilot who put most bombs inside a target from a given height.

The Zeppelins were flying bombs, filled with easily inflamed hydrogen. It took a really brave crew to man them, especially after the Englishman George Thomas Buckingham invented the incendiary bullet in 1915. This contained a filling of phosphorus guaranteed to ignite any Zeppelin hit from a distance of up to 350m (1,050ft). Attempts to improve the armament of the airships by exchanging some of the four, still watercooled, MG 08 of army type (two in the crew-/engine-gondola and one each in a special firing position on top of the airship, front and rear. These gave a wonderful field of fire, but were not made for gunners tending to vertigo …) for a better weapon.

## The 3.7cm *Luftschiff-Kanone* Krupp (the 1.5in Airship Gun)

This had been designed as a weapon with a one-shot kill capacity against the flimsy planes of those days, which whirred around the majestic Zeppelins like wasps molesting a whale. But when the airship gun was finished, so were the Zeppelins, the latter via the incendiary bullet. So in 1917 this gun was converted into an antiaircraft version by Krupp mounting it on a pivot. Since the short barrel only had L/14.5, amounting to 537mm (21.5in), a length that had been chosen to enable the handling of the gun inside the cramped airship gondola, it fired its solid shot with tracer at 350m/s muzzle velocity only (1,050ft/s). The rate of fire was satisfying, at 120 rounds per minute (this is obviously the cyclical rate of fire; the practical one was much lower as the empty ten-round clips had to be exchanged for full ones all the time) as was the ceiling of 2,200m (6,600ft). The gun disappeared from sight after 1918, even in literature.

More success and life expectancy favoured the other automatic cannon that armed the Zeppelins: the 2cm Becker *Kanone*. This has already been discussed in its other roles as antiaircraft, antitank and tank weapon.

*The 2cm Becker gun, mounted in the gondola of a Zeppelin airship. The high rate of fire – 400 rounds per minute – soon emptied the magazine holding only twelve rounds. On planes the 2cm Becker was mounted on the ring mount invented by Franz Schneider of Johannisthal.*

*Developments of 1916 and Beyond*

*The little-known 3.7cm* Luftschiff-Flak *by Krupp in its later role as an antiaircraft gun.*

**Aircraft Guns**

| Gun model | Calibre (in) | Weight Empl. (lb) | Tube Length (in) | Shell Weight | Muzzle Velocity (ft/sec) | Max. Range (ft) | Elevation/ Azimuth (degr.) | Remarks |
|---|---|---|---|---|---|---|---|---|
| 2cm Becker MK | 0.8 | 120 | 32 | 5oz | 1,500 | 7,500 | n/a | 400rpm (1) (2) |
| 2cm Ehrhardt MK | 0.8 | 130 | 40 | 5oz | 1,500 | 7,500 | n/a | 400rpm Rhein |
| 3.7cm Luftschiff-Flak | 1.5 | n/a | 19.2 | n/a | 1,050 | 6,600 | vert n/a | 120rpm Krupp |

*Remarks:* All three fired solid shot with tracers. (1) The rate of fire given here is from another source, but may indeed have been higher than in the antitank ground-version; (2) for firing upward over 45 degrees the recoil spring had to be further tensioned.

n/a = no data available.

## THE 21cm PARIS GUN

To end our tour of German guns of World War One, we will look at both the latest and the absolute highlight, the gun that is known under no less than three correct names: *Wilhelm-Geschütz* and *Paris-Kanone* in German literature, and Paris gun in English; and also under an erroneous name: *Dicke Bertha*, 'Fat Bertha', a name only appropriate when speaking of the 42cm *M-Gerät*, as we have learned. This gun shelled Paris from a distance of about 120km (80 miles), and could have reached 170km (106 miles) at the end of the war. It is not only famous for its range, incredible at that time, but also because almost all information about it disappeared right after the end of the war. The guns were cut up by oxy-acetylene torches, and the plans, calculations and all documents referring to it were destroyed by the German authorities. It has remained shrouded in secrecy ever since, and even its chief designer, the head of Krupp's construction bureau, Professor Rausenberger, of 42cm-mortar fame, was forbidden to publish his memoirs in 1926, because they would have referred to his work on that gun. It was a tremendous achievement in ballistics and gunmaking, yet also a tremendous failure in the military field. It was seventy years before two scientists, Canadian Dr Bull and US citizen Dr Murphy, were able to reconstruct the gun, using modern computers and based on the still-unpublished Rausenberger manuscript which they received from his widow. Dr Bull has since died; he was shot in Paris. Parts of the ultra-long-range gun he was working on then (the Babylon, for Saddam Hussein) may be seen at the artillery collection exhibited at Fort Nelson in Portsmouth.

In August 1914 the forts of Lüttich/Liège had been smashed by heavy German artillery, mainly by mortars of between 21cm and 42cm calibre (8–16in), the latter on a wheeled carriage. This was named *Dicke Bertha* by the grateful infantry that did not have to risk their lives by storming the forts, after the first name of the wife of the maker, Dr Krupp von Bohlen und Halbach, who had married the only living Krupp descendant, adding the famous name to his with permission of the Kaiser in order to keep it alive.

At the same time the German army kept advancing towards Paris, the capital of the enemy, which had been continually fortified from 1831 onwards, starting with four redoubts built from earth in this year, over fifteen forts and batteries built between 1841–46, another seventeen forts and thirty-four batteries built using the plans of General Sère de Rivière between 1874–1881, and in 1911 a new circle of forts 200km (125 miles) long was planned. In World War One these plans were replaced, because of their high cost, by field fortifications built in concrete.

When war started therefore, Paris was a *camp retranchée*, a fortified camp. This was in fear of the thrust of the German army advancing inexorably. On 28 August the Army of Generaloberst von Kluck was close enough for the Kaiser to thank him by radio message, saying that '… after decisive blows against Belgians, Britishers and Frenchmen the First army now in its rush towards victory is approaching the heart of France.' This message was intercepted by the French Intelligence and was not liked. On 2 September the French government left Paris for Bordeaux, but on the night of 3 September the 1st German Army was ordered by the OHL to 'push the French army to the south east, away from Paris'. Paris was not attacked by the Germans; the battle of the Marne started and then developed into the war of positions. Paris was never hit by a single shell fired by German artillery.

As the war went on, German air raids against the enemy hinterland were more and more endangered by the antiaircraft fire of the Allies, from both London as well as Paris. Thus, on the night of 22–23 March 1918 everyone had been sleeping peacefully in Paris, until roused at 07.20 by an explosion on the quay of the Seine. No one was hurt then, but at 07.45 another explosion, this time on the Boulevard de Strasbourg, killed seven people and wounded another thirteen. What had happened? The obvious explanation was bombing by enemy planes, but no one had noticed one or heard an engine. Then a new suspicion was finally proven after closer examination of the fragments: a German gun had fired them. But the front-lines had stopped 100km away long ago, and no gun could fire over that distance. So the hunt was on, with soldiers, gendarmes and planes all looking for where a German gun might be emplaced in a hidden position. The press, not wanting to miss this wonderful chance of selling extra editions, expounded on various theories for want of exact information. The *New York Times* wrote in their Sunday edition on 24 March of this sensational happening, that:

> … the officials in Washington do not believe in the existence of a gun capable of firing more than 62 miles. The shots may have been fired by giant airplanes bearing 9.5in howitzers. Others believe that the shots were fired by a British or French gun, manned by mutineers. There may also have been a giant centrifugue, throwing these shells.

Contrary to Washington opinion, however, there was such a gun: the German *Wilhelm-Geschütz*, the Wilhelm-gun, named after the Kaiser. This is the story of how it came into existence.

We heard earlier about the German *schwerstes Flachfeuer*, the super-heavy low-angle fire, for which naval tubes of mostly 38cm (15in) calibre, destined for never finished warships, had been mounted on makeshift *Bettungsschiessgerüste*, bedding firing racks, firing 27km (16.9 miles) and designed by Krupp. One of them had been emplaced near to each French fortress along the German–French border. One had also been planned

*Developments of 1916 and Beyond*

*When the first shells fell on Paris in 1918, the French media had only a vague idea of what the gun firing at them looked like.*

for use against Belgian Antwerpen near Mechelen, south of Antwerpen, in October 1914, but remained unfinished when the fortress was taken. Firing at fortified Paris was an obvious development of this idea. In the 1870–71 war, Paris had been the target of no less than 110,000 shells; now the heavy 42cm mortars of *Gamma-* and *M-Gerät*-type would need less rounds to achieve the same success.

Naval guns had increased in calibre from 28cm (11in) to 30.5cm (12in) in 1906. The next upgunning was the topic of a conference of the *Reichsmarinamt*, the Imperial Navy Office, on 1 September 1911. It had still to be decided whether this next step was to result in guns of 35cm (14in) or the British battleship calibre of 38cm (15in). The problem was that the life expectancy of big naval guns dropped with every round fired, and the range lowered by 10m (30ft) for each round, too. Krupp felt that the British would not exceed 40cm (16in) – and they were right. In the end, Admiral Tirpitz decided to propose this calibre of 16in to the Kaiser. On 6 January 1912 the Emperor ordered that the battleships constructed in 1913 should be armed with 38cm (15in) L/45 guns.

The leapfrogging did not end there. On 26 October the Kaiser responded to larger British naval calibres with an order to design a ship with 42cm (16.8in) guns, and in April 1918 the admiralty even asked for designs of 50cm (20in) twin gun turrets, declined by the naval weapon department because Krupp was already too busy designing and producing other guns. Supplying the artillery for the army and the navy was good business for Krupp. The twenty-six guns of all calibres, from 8.8cm up to 38cm for one of the two new battleships, *Bayern* and *Baden*, alone cost 23.4 million mark, almost half of the 58.3 million for one ship. One 38cm gun in 1915 was 451,700 for the tube plus 105,500 for the cradle. If these guns were to fire from emplacements on the ground, the mounting and the concrete bedding would be another half million – one million altogether for a 35cm or 38cm gun. It would be served not only by sailors – specially trained naval gunners – but also by well-trained technical personnel from the Krupp factory.

These naval guns for land targets were now ordered from Krupp; the OHL started by demanding a 35.5cm gun with which to fire at Dover from an emplacement at Cap Gris Nez near Calais. (This was also where, after 1940, the 3 × 40.6cm (16in) Lindemann-battery and the 4 × 38cm Todt-battery were constructed, the first now gone in a mountain of chalk with the formation of the Channel Tunnel.) Since the war took an unexpected turn, this gun remained where it was and later found other targets in France. The navy could not do the job of shelling Dover from this distance, since the embrasures in the twin turrets gave an insufficient elevation of 20 degrees, resulting in a range of only 23km (14.4 miles). No experience had yet been gathered with these guns on land firing with higher elevations.

## Developments of 1916 and Beyond

The ammunition side had been investigated before the war, by experimenting with shells with a larger ogive, the nose of the shell. If this was made more pointed by having the radius of the ogive increased from the usual two to three times the calibre, to ten times the calibre, the calculated range for the 745kg (1,560lb) 38cm shell was raised to 28km (17.5 miles) at a muzzle velocity of 800m/s (2,400ft/s); and the faster-starting (900m/s:2,700ft/s) 35.5cm shell of 535kg (1,070lb) reached out to 39km (24.4 miles).

Experiments at Meppen with the 35.5cm L/52.2 gun showed surprising results: geometry said that the longest range is obtained if you fire with an elevation of 45 degrees. Actual test firing disputed this. The shells went furthest if fired at a higher elevation of about 52 degrees. The reason was found to be the air. As we know, the density of the atmosphere decreases from the ground upwards, because the weight of the molecules compressing it becomes less and less. It is also known that the air slows down everything moving through it, the infamous air resistance which increases by the square of velocity: double the speed means four times the resistance. So if a shell is fired through thinner atmosphere it will encounter less resistance on its way, and stay faster and go further. By firing at elevations over 45 degrees the shell came into thinner atmosphere sooner, losing less velocity and gaining in range. Fifty-two degrees seemed to be the optimum compromise between elevation and range, so this was chosen.

By 1918 the 35.5cm guns were firing at 62km (38.8 miles) and the heavier 38cm shells reached 48km (30 miles). There had been other solutions possible. One would have been to use the subcalibre-discarding sabot shell, which today is fired from most tanks as armour-piercing KE-round (kinetic energy). As a matter of fact, one of the ballistic experts working with Rausenberger, von Eberhard, suggested such a shell. This idea was refused for unknown reasons.

Let us look at who came up with the idea of such a 100km gun. True to the experience that success has a thousand fathers and failure none, more than one man later admitted modestly that it had been his idea.

One of the names dropped was General Ludendorff of the OHL. Another was a Colonel Bauer, who wrote that already in 1908, when he had suggested the design of the 42cm mortar, he had had to fight opposition within the artillery and at Krupp too, except for Rausenberger. The same happened again when he later demanded long-range guns. Even Kaiser Wilhelm, in his memoirs, also remarked that he, the Kaiser, had had to fight objections against the real long-range guns, especially from the artillery. (Having belonged to the ordnance corps, I must admire the wisdom of the artillery officers, who had already recognized that these super-long-range guns were a waste of money, time, gunmaking capacity and powder, resulting in the end in a strategic mistake without any gain, military or political. It was a forerunner of the 80cm K (E) DORA of World War Two, of the same calibre as a mistake. But as a gun it was of course a wonderful creation, far ahead of its time, a very costly toy for scientists and the military.) It seems that Colonel Bauer had contacted Rausenberger in November 1916 about using a big calibre 35cm or 38cm gun and inserting a subcalibre tube of 21cm into this. This would save a lot of time, since the 35cm and 38cm guns were already in the field, not only on ships, but also on the ground.

And Rausenberger wrote that in the autumn of 1916 the *Reichsmarinamt* (RMA), the Naval Office, told Krupp that a certain cruiser, *Ersatz Freya*, which means that this was planned to replace the old *Freya* but had not been named yet (it was to be baptized *Prinz Eitel Friedrich*), was to be finished only in 1918. Therefore the nine 35cm guns for this new cruiser could wait. But these guns were already almost finished except for the rifling process. So Rausenberger proposed to insert long 21cm (8in) tubes into these 35cm guns – a few months' work – which would leave them intact enough to be converted into 35cm guns again, if the need arose. Designing, building and testing completely new tubes would have taken about another eighteen months. Rausenberger also realized that firing a 21cm (8in) shell of only 100kg (220lb), thick-walled because of the high pressure in the tube caused by the tremendous amount of powder necessary for such a range, would bear a charge of only 8–10kg (17–22lb) explosive. This was certainly not sufficient to destroy military targets such as harbours, depots or railway junctions. The target would have to be an important city like Paris because of the moral effect of such a shelling on the population. So he told his friend Bauer, and Bauer asked Ludendorff, who gladly agreed to this idea and gave permission.

Bauer remembers this differently; his targets were the French harbours, since London was too far away. Paris came later as a target, during the spring offensive of 1918. Until then, both Hindenburg and the Kaiser asked Bauer about the development of the long-range gun, which made only slow progress.

Up to now the idea had been for a gun firing 100km (62.5 miles), when, according to Rausenberger, in February 1917 a letter from the OHL and signed by Ludendorff told Krupp that '… due to new considerations the 21/35 must have a range of 120km (75 miles). Please continue your work on this basis.' The new considerations were the German front-lines, which had been pushed back by the Allies in the meantime.

Rausenberger continued with his fresh calculations. The extra 20km (12.5 miles) may not seem much to a layman, but for us the problems caused by trying to fire at 100km increased tremendously. Firing at 50 degrees elevation, the 100kg 21cm shell for 100km had needed 1,500m/s (4,500ft/s) muzzle velocity and taken 120kg (252lb) of powder to reach this, resulting in a maximum pressure of 3,600atm (52,200psi). Now these numbers rose to 1,610m/s (4,830ft/s), 200kg (440lb) powder, and 4,000atm (58,000psi). And the shell had to become even stronger-walled because of the higher pressure, taking only 7kg (14.7lb) explosive now. Note that increasing the range by 20 per cent required 70 per cent more propellant and gave 12.5 per cent less explosive payload. The powder had posed enough problems already. The navy used a powder of their own, best suited for the heavy naval guns, the *Röhrenpulver* C/12, the tubular powder, introduced in 1912, which held 25 per cent of nitroglycerine and 70 per cent of nitrocellulose in its macaroni-like sticks of 820mm (32.8in) length. For the Paris gun the sticks of the 200kg charge had to be 1,230mm (49.2in) long. The gunners of the army used to accuse their naval comrades of wasting powder. This is understandable if you compare the 38cm naval gun firing a shell with only 33kg (70lb) explosive, using 205kg (430lb) powder to do this, with the 42cm mortar dispatching 100kg (220lb) explosive with only 75kg (158lb) powder.

Ballistics were also unfriendly. Air pressure and temperature, wind force and direction; even the rotation of the earth during shell travel, had to be considered in exterior ballistics. The imminent danger was of shells tilting because of insufficient gyroscopic stability, and falling short. The interior ballistics also caused headaches: wear of the heavy calibre tubes was such that each shell fired shortened the range by about 10m (30ft). Erosion inside the chamber lengthened this for each round by 15mm (0.6in). The temperature of the powder before firing affects its rate of burning and maximum pressure, with the muzzle velocity increasing for one promille with each added degree of centigrade, resulting in greater dispersion.

But the 21cm *Wilhelm-Geschütz* was now ready for testing. So what did it look like? Less than a dozen photographs of this gun have survived and are well known to enthusiasts. These, along with the description given here, provide a general idea of what the gun looked like (but of course no details were available on the design or the materials used).

Let us start with the actual gun, the weapon in the mount. Since the order from the OHL via the navy arrived at Krupp at the end of 1916, manufacture of the first batch of the 21cm gun may have started at the beginning of 1917. A second batch followed later in March, when experiences with the first batch were available, and new calculations had been finished. (Remember that in those days, computers did not produce the answer within milliseconds. Partial result after partial result was recorded by hand on sheafs of paper, sometimes taking months.) Work continued round the clock in the Krupp factory: in drawing offices, construction shops, laboratories,

*Developments of 1916 and Beyond*

*This is what the 21cm gun, called* Wilhelm-Geschütz *or* Paris-Geschütz, *the Paris gun, really looked like.*

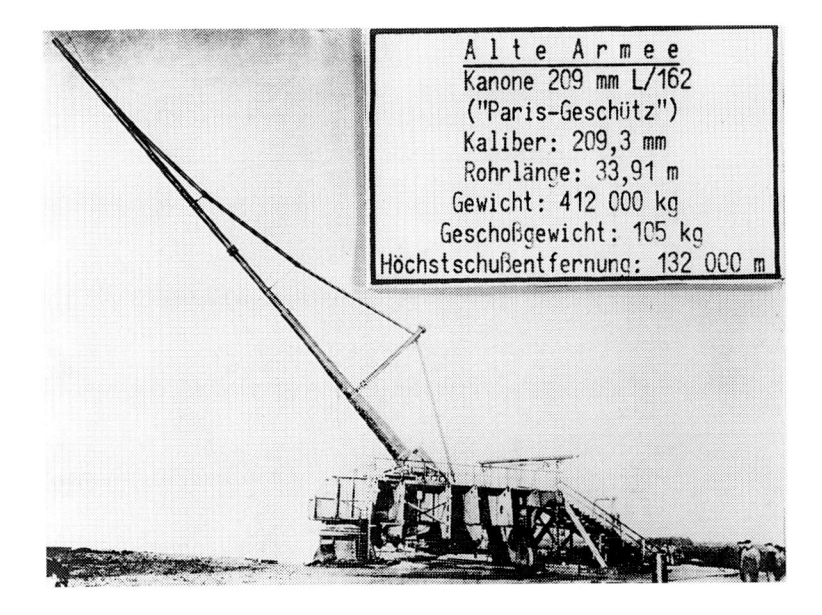

```
Alte Armee
Kanone 209 mm L/162
("Paris-Geschütz")
Kaliber: 209,3 mm
Rohrlänge: 33,91 m
Gewicht: 412 000 kg
Geschoßgewicht: 105 kg
Höchstschußentfernung: 132 000 m
```

foundries and machine shops. All men laboured, not for money, but for their soldiers, lovingly referred to as *Unsere Jungs*, our boys. Material that would stand the hitherto unheard of strains when firing this way was searched for and carefully tested. We do not know the alloy, since literally not a single gram of *Wilhelm-Geschütz* has been found since the war; every single scrap being first cut up and then returned to the blast furnace.

To speed things up, the guns were built using existing guns as a base. This started with the 35cm SK (*Schnelladekanone*, rapid-loading gun; and later, when all guns were of this type, interpreted to mean *Schiffskanone*, ships gun) L/45. Of these, the tubes numbered 11–18 were used. Later, in April 1918 38cm tubes SK L/45 were also used for this purpose, to guarantee the furnishing of two *Wilhelm-Geschütze* a month for the front. The 21cm liner of 21m (63ft) length was inserted from the rear. When worn out after a certain number of rounds (the exact number was still to be found during the tests), these 21cm liners were to be removed and then rerifled in two stages to larger calibres: first to 224mm (8.96in) and then again to 238mm (9.5in). The twist of the rifling was constant, one turn in thirty-five calibres. This had been

a matter of concern for Rausenberger, who feared that this might be too much to keep the shell following the curve of the trajectory, causing it to stay at the same angle of departure all the time to its impact, which then would not be with its nose but its base. In this case, the higher drag would also shorten range. The shell rotated 200 times a second when leaving the muzzle.

This 21cm liner was made in three parts for ease of manufacture, and the sections placed end to end to end. Another problem of production was the fact that Krupp only had machines for rifling tubes up to 18m (54ft) long. Calculations had shown that for the 1,610m/s muzzle velocity needed for the desired range, the shell had to travel inside the tube for at least 24m (72ft). The missing 6m (18ft) was supplied by a smooth tube, which could be mounted in the emplacement, known as the *Tüte*, the cone. Cones of 9m and 12m (27–36ft) length were also tested. The continuing calculations with differently shaped shells later resulted in guns firing at even longer ranges: the *Wilhelm-Geschütz* would have reached 142km (88.8 miles), enough to shell London from Calais; and a new 30.5cm (12in) *Weitschussrohr*, long-range tube, was to fire a shell of 300kg at 1,700m/s (5,100ft/s) over a range of 170km (106

*Developments of 1916 and Beyond*

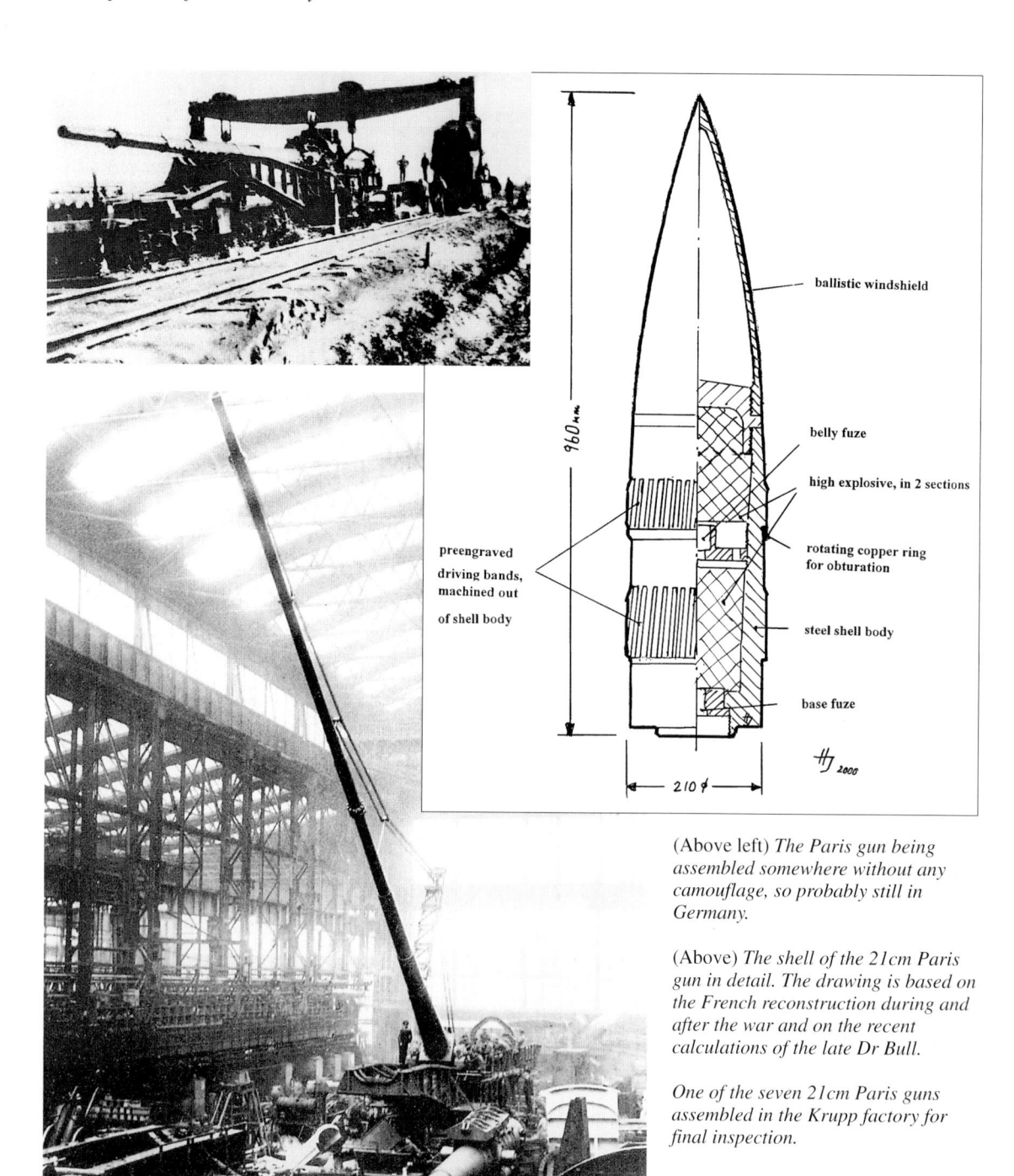

(Above left) *The Paris gun being assembled somewhere without any camouflage, so probably still in Germany.*

(Above) *The shell of the 21cm Paris gun in detail. The drawing is based on the French reconstruction during and after the war and on the recent calculations of the late Dr Bull.*

*One of the seven 21cm Paris guns assembled in the Krupp factory for final inspection.*

170

*Developments of 1916 and Beyond*

miles). (In World War Two the Germans resurrected this idea in 1944 and started building a huge battery with some fifty multichambered tubes into a chalk hill south of Boulogne, which was to fire salvoes of 15cm (6in) fin-stabilized subcalibre shells of PPG-type, *Peenemünder Pfeilgeschoss*, towards London. It was bombed while still unfinished.)

The liner gun was loaded with the shell first and then the powder charge of 190–200kg (400–420lb): the 50.5kg (106lb) or more of the *Zusatzkartusche*, the additional charge, then a similar *Vorkartusche*, front charge, of 75kg (158lb), and finally the 70kg (147lb) of the *Hauptkartusche*, main charge, contained in the regular brass case of the 28cm navy gun Kanone L/45. The 70kg of the main charge and the 75kg of the front charge were always the same; only the additional charge varied from 50.5kg upwards. Other authorities of German artillery later wrote about '195–196kg' (390–392lb). We also know that the charge had to be increased after each round to make up for the wear of the tube; one authority (US Lt Col Miller) stating this to have been 1kg (2.1lb) for every 100mm (4in) advance in the seating of the round, with an advance of 1.8m (6ft) altogether during the normal barrel life of forty-eight rounds (Rausenberger had implied seventy to eighty rounds). That would give an advance of 37.5mm at each round, or 1kg more powder every third round, with the total increasing by 16kg (33.5lb) from the first to the last round. (All data given varies to a limited degree according to the different authors. This stems from the strict secrecy maintained by the Germans even after World War One, combined with deliberate misinformation about all vital things about this gun. Ballisticians will enjoy the final calculations of Dr Bull on this matter in his book.)

The gun is described as being 34m (102ft) long, 1m (3ft) in diameter, with the walls of the breech being 40cm (16in) thick, tapering towards the muzzle. Its weight was 200 tons. The long tube had to be safeguarded against droop (bending under its own weight), which would have amounted to 90mm (3.6in). This was done through the addition of a *Spannwerk*, a bracing, onto the upper side of the tube. This was formed by beams vertical to the

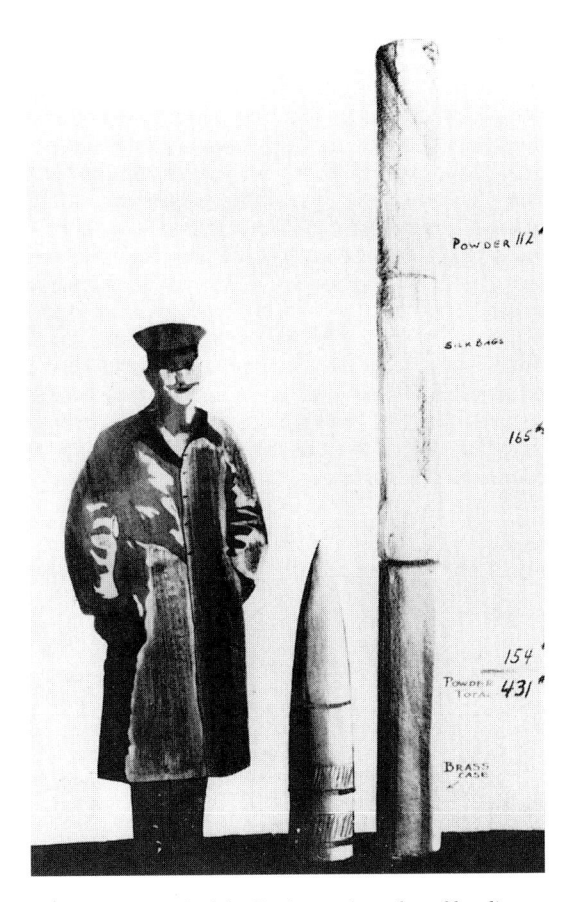

*The whole round of the Paris gun in order of loading: 21cm HE shell, first powder charge of 111lb, second powder charge of 165lb – both in silk bags – and finally the brass cartridge case of the 28cm SK, filled with another 154lb; altogether a charge of 430lb of powder to fire the shell over a distance of 82 miles.*

axis of the bore, connected at their tops by a cable running to the muzzle. Tensioning this cable pulled the muzzle upwards, until the axis of the bore was absolutely straight when observed through a telescope from the inside of the breech.

Test firing had to be done at another firing range, Meppen not being long enough to permit ranges of 50km (31 miles) and more. In December 1916 the navy furnished their new range of Altenwalde on the coast near Cuxhaven, together with a mixed crew consisting of both sailors of the navy and soldiers of

171

## Developments of 1916 and Beyond

*The Paris gun firing on the naval range of Altenwalde.*

(Below) *The Paris gun during testing on the Altenwalde range.*

(Bottom) *The ganz geheime, completely secret, firing table for the 21cm Paris gun.*

the reserve: one officer, two noncommissioned officers and 84 men altogether, only 36 of whom were classed as fit for war. There the *Wilhelm-Geschütz* fired its first rounds, in a westerly direction over the islands on 23 July 1917. Colonel Bauer represented the OHL then and later reported that:

> Firing results seemed to disprove all former calculations . . . Dispersion was terrific, with rounds calculated to impact in 70km (43.8 miles) distance, landed at 50 and 80km (31–50 miles). We did not find out the reason for this, since we could not recover any of these shells at first . . . Since the first rounds had shown the tube to stand the strain of firing, we did not deem it necessary to take cover inside the shelters later. This almost proved to be a fatal error, when suddenly the front part of the tube went over our heads with terrible noise. This ended the firing for that day.

Rausenberger placed the blame on driving band replacements shearing off, the wrong sort of powder (which had to be stored at over +15°C), and the windshields of the shell flying off. The most serious problem lay in exactly centering the shells. For this, instead of normal driving bands made of copper, the body of the shell had been machined so that a sort of pre-engraved rifling was left in two sections, at the front and rear of the projectile. Behind these steel projections copper rings were inserted, their diameter depending on the wear of the tube, growing with each round. Rings made of soft material, tin

## Developments of 1916 and Beyond

and asbestos, sealed the space between shell and tube so that no gas could escape over the projectile toward the muzzle and be lost for propulsion. With this, a longer *Tüte*, the smoothbore cone in front of the muzzle of the rifled part of the long 21cm tube, of 12m (36ft) was installed.

The 100km (62.5 miles) range was passed for the first time on 20 November 1917, and 126km (78.8 miles) on 30 January 1918. These rounds were fired with 50 degrees elevation, a muzzle velocity of 1,645m/s (4,935ft/s) and a pressure of 4,200atm (60,900psi). The shell took 182 seconds to travel to the point of impact, descending under an angle of 56 degrees (this angle is always greater because of drag, than the one of departure, the elevation of the gun). The zenith of the trajectory lay at a height of 40km (25 miles), thus putting a man-made object into the stratosphere for the first time.

The guncrew was made up of gunners of the naval artillery, which manned all heavy German guns in World War One. These came from the *Marineartillerie-Regiment*, the naval artillery regiment, at Kiel. They were sent to Crépy near Laon, where in the neighbouring woods work had already started on the concrete bedding. This swallowed 100 tons of cement, 200 tons of gravel and 2.5 tons of wire netting. Camouflage of the site was most important right from the beginning, as British planes were nosing around all the time. Then the gun and the mount were put on the railway at Essen and sent on to Crépy. Since these first guns were not railway guns, but used the rail tracks solely for transport, the 250-ton gun carriage was lowered onto the 300-ton concrete pedestal, already causing a high ground pressure, which would be increased tremendously by the forces of recoil. The firing was to be under the command of Vice-Admiral Max Rogge (made famous by the 38cm Max railway gun), Chief of the Naval Ordnance. Rausenberger, of course, was also present.

The gun was cranked by hand into the calculated azimuth, then the axis of the bore verified again. Now the shell was rolled to the gun on its cart by hand. Each shell bore a number, for each was different, made with regard to the calculated wear on the tube and compensating for this with two larger rotating copper bands for obturation, below the pre-engraved rifling machined out of the steel of the shell body in these two places. Each one was also armed by two fuses – one forward fuse, seated around the middle of the shell on a firm support so it would not get a soft impact sitting on the explosive, and an aft fuse at the base. The shell had an aerodynamic windshield and weighed 106kg (220lb), carrying only a ridiculous-seeming payload of 7kg (14.7lb) trotyl explosive, no more than a 155mm shell of today. It was first carefully inserted until the raised ribs of the shell slid into the grooves of the rifling, and then rammed home in the horizontal loading position of the tube by four soldiers using the rammer (a long pole). Like later rounds, the correct seating was determined by the ramming force necessary, thus ensuring that even chambers lengthened by long firing still had the shell seated correctly.

Now the powder, kept warm at about +15°C since the day it had left the factory in a heated armoured cartridge car, and watched over in its storage bunker in the battery by an officer who was an ammunition expert, was rolled to the gun and loaded into the chamber. The exact weight of the charge had been calculated a short time before, on the meteoric data of air temperature, air density, wind direction, and so on, but only for the gun site, not for the long way to the target. All three parts of the charge went in, not rammed the hard way like an ordinary shell to be seated properly, which in a gun means that the lands of the rifling have to 'bite' into the copper of the driving band, but softly shoved into the chamber, so neither the enclosing silk bags nor the long tubes of the powder inside would be damaged. The heavy breechblock closed, 1,850kg (3,885lb) of a *Schubkurbelkeilverschluss*, the sort of breechblock that the 7.7cm FK 96 n/A shared with the 42cm *M-Gerät*. The gun was elevated, again by hand, a really exhausting job for the sixteen soldiers working the long crank, in spite of the counterweights fastened onto the top of the breech rim for equilibrating the weight of the tube. These men had to be spelled often.

## Developments of 1916 and Beyond

The muzzle of the gun appeared somewhere among the artificial part of the woods, created by poles and wires holding all the different camouflage nettings in place, painted in spots of brown, green and dark red; this was the *Bunttarnung*, the multi-coloured camouflage of those days. The crew took cover inside a shelter. Only the chief gunner remained in the open, his hand on the firing switch (the gun could also be fired manually by pulling a long lanyard). Now the gun was ready to fire. But it waited, letting the other guns assembled around its position go first, for the *Geräuschtarnung*, the noise camouflage. By now the art of acquisition of enemy batteries was well developed, both flash ranging and sound ranging endangering the life of a gunner, who was used to firing peacefully after the Longfellow-principle ('I shot an arrow in the air, it fell down I know not where'), but now had to take care that not one of the dangerous British Tucker microphones fielded in 1916 picked up his gun signature. So from the command post of Admiral Rogge, who of course was linked to the OHL, the supreme army command of the Kaiser, came the order to some 30 covering batteries: '*Feuer frei*', open fire. And they did, with all sort of calibres, from field guns to heavy railway artillery. Then the OHL gave the order '*Feuer frei*' to the *Wilhelm Geschütz* by telephone. The order rolled from the man on the switchboard to the commanding officer, and then on to the chief gunner, who repeated for confirmation '*Feuer*' and flipped the firing switch, closing an electric circuit.

And then, for the first time, the 21cm *Wilhelm Geschütz* joined in this chorus of more than 100 guns. The noise of the shot was not as loud as expected (the super-long tube keeping the muzzle well away from the ears of the gunners), the gun recoiled for up to 1,350mm (50in) only, as due to the light shell recoil was less than with a 38cm Max gun. The large black cloud created by this smokeless powder (ironically, the cloud would have been white in the case of blackpowder) disappeared, but the slender tube quivered like a fishing rod for minutes. Rausenberger, like all the others, looked at the second hand on his watch, measuring the time of flight towards Paris.

Three minutes after firing, the shell impacted on the Seine *quai*, by accident even hitting a military target: a detachment of sappers marching there. Two soldiers were killed and several wounded, a pure accident when firing at a target like Paris of 12km (7.5 miles) diameter. It has been pointed out before that even today it is considered something of a miracle that the Wilhelm gun hit Paris at all at this distance of 120km. And this with ever-changing charges and shells.

But how did the gunners know how the round had performed; whether it had held the range, and whether the shell had stayed within a certain circle of dispersion? The answer was provided by a small blind passenger hiding inside the gun when it fired: the *Messei*, 'gauging egg', the egg-shaped copper crusher gauge. This invention, dating back to the previous century, was a fist-sized steel cylinder that recorded the maximum pressure of the gas inside the tube, by letting this push a plunger onto a copper cylinder of known malleability, compressing this in length. Two of these gauges – so their results could later be averaged – were inserted in special niches of the chamber wall before the main charge was loaded. They did not leave the tube with the shell either, but stayed inside and were later removed after the breech was opened again. (When used in other guns the eggs were simply loaded inside the propellant and either flew for a few metres or remained inside the tube, depending on elevation.) The crushed copper tube, when removed from the egg, was then measured carefully with a micrometer gauge and the new, shorter length recorded. A table then revealed the maximum gas pressure of this shot. Experts will object that it would have been better to measure the muzzle velocity of the shell. But until World War Two this was done by firing through two wire mesh screens, with the gun in horizontal elevation. The shell set off an electric clock when penetrating the first, stopping it again by hitting the second. From the time needed to cover the known distance between the two screens, the speed was calculated. This was obviously not suitable for firing in the field (but today's method of carrying the coils of the velocity measuring equipment is: just look at the

*Developments of 1916 and Beyond*

35mm tubes of the twin guns of the German antiaircraft tank GEPARD).

So Krupp had tables recorded during firing, at what gas pressure the different velocities had been reached, and then took the pressure as an analogue figure for the speed; simple but effective. The copper crusher gauge is used even today to find out if the pressure is still in safe limits. The other method – drilling holes into the tube and fitting piezoquarz elements for pressure recording – is obviously only practical for laboratory work, not for guns firing in anger.

When the gun had been cranked down again – other sources state that laying the gun was done by electric motors – to the horizontal loading position, the breech was opened. Normally the the cartridge case is extracted and ejected by the extractors, to be later resized in a depot, getting a new primer screw and being filled again with powder. But not here, as the case was reduced to a slag of molten brass by the burning of so much powder.

Not all of the firing happened so smoothly. There was, of course, enemy counter-battery fire once the position at Crépy had been found out by sound ranging. Combatting the three guns was the job of French 30.5cm (12in) railway guns. They came close and also hit part of one crew, but did not destroy one of the guns. This is said to have happened (and furiously denied by others) on the morning of 23 March 1918. When gun number three fired, a blinding flash erupted from the breech, which was torn off completely. Some of the crew were tossed far away into the woods (they had not left the gun to take cover, since 'nothing had happened yet', the usual astonished argument in such cases). Five were killed and 12 seriously wounded.

One shot cost about 10,000 mark, according to General Ludendorff, or four to five times that of a French 30.5cm gun. Was it worth this? For a gun, the utility was judged by calculating the ratio of weight of the gun and that of all the shells fired during its life. This figure was 1:600 for the 15cm sFH; 1:100 for the 42cm *M-Gerät*; and for the 21cm Wilhelm gun 1:0.09, less than one thousandth of the number for the *M-Gerät*.

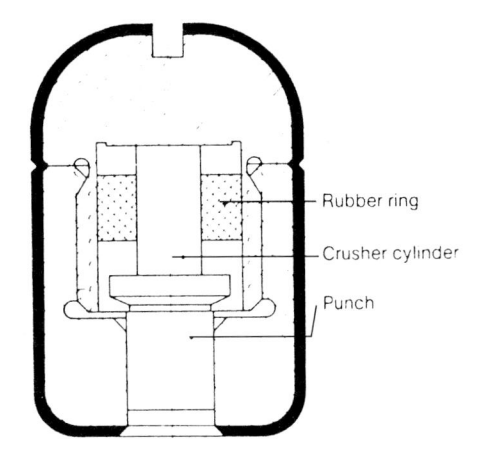

*This little bird told the gunners of the Paris gun where their shell had hit. Egg-sized and -shaped, two of them sat in the niches of the gun's chamber, recording the peak gas pressure during firing. When removed after the shot, the recorded pressure was the basis for calculating the muzzle velocity of the shell, and from this the range.*

So what of the military, psychological and political aspects? Back to Paris and the first rounds impacting there on the morning of 23 March. Three guns had been installed at Crépy. On the 23rd, two of them fired, about 900m (2,700ft) apart, one after the other. A few days later the third joined them. The effect of the fire could not be observed, neither by forward observers nor even by planes. It was pure unobserved fire. Some days later the effect could be read in the French newspapers. It is also said that a German spy living in Paris reported it via Switzerland. Actually, there was not much to report. The 343 rounds fired at Paris (161 rounds out of four 209mm (8in) tubes and 182 rounds out of two rebored 232mm (9.28in) tubes) between March and August 1918 are said to have killed 256 people – primarily due to the accidental hitting of two full churches – and wounded 620, causing mourning, fright, and fury among the citizens of Paris. Some left the city, but most stayed. The public was not disturbed enough to create civil unrest; nor was the military head and nerve centre impaired. They were of less effect than the Anglo-

175

## Developments of 1916 and Beyond

American bombings of German cities in World War Two in detracting from the staying power of the population. Still, after the war neutral observers agreed that at this time the Wilhelm gun had been the only German weapon for this range.

After the war these guns vanished into thin air. The hardware was cut up and turned into new products, all documents were eradicated and strong measures prevented all living witnesses from giving out any details. Thus the secret of the Paris gun was kept and is lost, in spite of all later reconstructions.

Most other nations were eager to discover it. The French actually tried to build a long-range gun of their own: the 210mm gun M 23. It was of the same calibre, but unlike the German model it fired from a railway mount, the old sliding mount of the 340mm gun M 1912. The M 23 had a tube of only L/110, compared to the German L/162, and fired a shell of 108kg (237lb) with a muzzle velocity of 1,450m/s (4,350ft/s) out to 120km (75 miles), not quite up to the standard of the old master, so only one was produced.

When in 1934 the German *Heereswaffenamt*, the arms office, wanted a long-range gun, there seems to have been sufficient knowledge after all still stored in the heads of the surviving generation that had constructed the Paris Gun (Rausenberger had died at 59 years of age in 1926) to have a modern version built. This was called the K 12 (E), showing that it could fire (10 times the number given) 10 × 12, 120km (75 miles), and was an '(E)', *Eisenbahngeschütz*, railway gun. Test firing started in November 1937 at Rügenwalde, the gun was finished in 1938 and issued to the troop as 21cm *Kanone* 12 V (E), the V for *Versuch*, experimental version. This was followed by another one, this time the 21cm *Kanone* 12 N (E), the N for *Normalform*, the standard version. Both were built in a similar way to the old *Wilhelmgeschütze* – bracing, dimensions and all – and also fired a shell of 107.5kg (226lb) out to ranges up to 115km (72 miles) with the same negligible effect. But they were genuine railway guns now, firing from curves of the track, turntables, or track parts crossing at 90 degrees. They were deployed in 1940 and during the war on the French coast of the Channel, firing seventy-two rounds altogether until 1941, most of them innocent of any damage to British property. But they served as ballistic laboratories, assisting in the further development of long-range shells. Therefore it was also to their merit that the 28cm K 5 (E) later extended its range from the 50km (31 miles) indicated by the '5', to 86.5km (54 miles), firing a full calibre rocket assisted shell, and even to 160km (100 miles) with the subcalibre sabot shell, fired from the later K5 version, the smoothbore K5 *glatt*. This showed where the future lay: Rausenberger should have listened to Eberhardt after all.

Today, the rocket assisted projectile (RAP) with its too great dispersion is no longer the hope of the artillery. 'Base bleed', where a tracer-like cylinder screwed to the base of the shell generates gas to even out the turbulences there and reduce drag, seems to be the new answer to the age-old quest for longer range.

### The Paris Gun

| Gun model | Calibre (in) | Weight Empl. (tons) | Tube Length (in) | Shell Weight (lb) | Muzzle Velocity (ft/sec) | Max. Range (ft) | Elevation/ Azimuth (degr.) |
|---|---|---|---|---|---|---|---|
| 21cm Wilhelm-K | 8.4/9.3 | 404/412 | 1,361 | 221/252 | 4,800 | 396,000 | +55/360 |

*Remarks:* used carriage of 38cm L/45 railway gun; calibre rebored from 21cm to 23cm; weights for 21/23cm version; transport by railway; firing from bedding; seven guns built; all disappeared at the end of the war.

# 4 German Artillery in Battle

For the artillery on both sides, World War One was a period of continuous fighting, without interlude. Of these four years we have looked at the instances of German artillery combating enemies such as tanks or aircraft, and special guns such as those in fortifications, both on the coast and on the borders. Now we shall look at the fighting in those theatres of war likely to be of particular interest to British readers. These include the battle of the Dardanelles, the fighting in Africa, and the two battles of Verdun.

## THE BATTLE OF THE DARDANELLES, 1915

The plan of First Sea Lord Churchill had been to force the Dardanelles with a fleet, and then attack Constantinople/Istanbul and thus draw the hitherto neutral Balkan countries and hopefully also Greece into the Allies' camp. At the same time, their Russian allies, pressed hard by the Austrians, could be relieved through the Black Sea. Germany saw this danger, both for her Turkish ally and for her Austrian ally too, whose undefended south flank would then be attacked by new enemies.

In the spring of 1915, a mere 170 German soldiers were sent to the Dardanelles to reinforce the Turkish defences there, all of them old brick and earthwork fortifications built after the Russian–Turkish war of 1877–78, without armour or concrete. The guns were not much younger – a museum-like collection of all calibres and vintages. Only one battery of Fort Hamidié mounted modern Krupp guns of 35.5cm (14in) and was put under command of German Lieutenant Commander Wossidlo. The commanding German officer in the Dardanelles, Admiral von Usedom, had only the heavy field howitzers of Lieutenant Colonel Wehrle's artillery regiment as an active defence, deployed on both banks. These 15cm howitzers continually changed their emplacements, always in hidden positions in the mountains, and fired in the high-angle mode at the British and French warships, which were unable to see them. When the combined British and French fleet attempted to force the straits, it was stopped by the last of the ten minefield barrages closing them. Ironically, this had been laid by Turkish and German sailors, using mines they had 'borrowed' previously from Russian minefields in the north.

Then the army landed on the Gallipoli peninsula and tried to take both sides of the Dardanelles, before attacking the small Turkish fleet and opening the way for Russia for good. This failed after eight months of bloodshed with the British and French retreat from the Dardanelles at the end of 1915. According to Churchill, in his book *The World Crisis* (1953), it was not superiority of equipment that decided this action, but willpower. Official records show the following numbers of losses: 55,000 Turks killed, 166,000 losses altogether; 32,000 British killed, 120,000 losses altogether; 3,700 French killed, 23,300 losses altogether. But the worst thing was that now Russia was isolated, with terrible results. The laurels for the heroic defence of the straits have to go to the Turkish defenders, both the gunners, dying in the open emplacements at the side of their old coastal guns, and the hungry infantry men, storming against the invaders in the terrific fire of the modern British ship guns. But German field artillery also played a part.

## GENERAL VON LETTOW-VORBECK AND THE WAR IN AFRICA

Germany was a late entrant into the market for colonies, the ripest fruits having already been picked by various sea-going nations. (There was one exception. In the seventeenth century, the *grosse Kurfürst*, the great elector, Friedrich Wilhelm von Brandenburg (1640–1688), had tried to get a slice of the cake of trade with Africa for Brandenburg and himself. A former Huguenot *corsair*, pirate, helped him by founding a trading company similar to the British or Dutch East Indian ones. Friedrich even built a fortified town for its centre, which today is Fredericksburg in Ghana.) Further endeavours from German traders came much later, and at the beginning of the war the *Deutsche Reich*, the empire, held some areas of Africa. Until 1914 these had been protected by the *Schutztruppe*, the protective troops, more or less a branch of the services, against natives not recognizing the blessings they received from the white man. Now these troops were called upon to defend the German colonies against formerly friendly white neighbours of all nationalities, not only British, French and Belgian, but even the Portuguese, who could hardly claim they too had been attacked by the German army. The *Schutztruppe* did not have much in the way of artillery. A half dozen old 9cm C/73 guns of 1873 vintage had been thought sufficient to disperse any uprising. But they were of little use against the superior numbers of the enemies now rushing with modern arms to claim a piece of the terrain held by Germany in Africa.

Most of the land fell in a short time; only one area, *Deutsch Ostafrika*, German East Africa, held out for the entire four years and against an enemy twenty times its superior in number. This was thanks to the grand leadership of the commanding

*Members of the German* Schutztruppe, *the colonial forces in the African colonies, with an old 9cm C/73.*

*German Artillery in Battle*

*Except for a few of these C/73 guns, at the outbreak of* World War One *the* Schutztruppe *only had the 3.7cm Maxim automatic cannon, lying in the carriage of the old 9cm gun, for artillery: this weapon was used as a landing gun by the* Marineinfantrie, *the naval infantry.*

*(Below) When the German* kleine Kreuzer, *light cruiser,* Königsberg *was sunk by British monitors in an African river in 1915, her 10cm guns of naval-type* Schirmlafette *were salvaged by the* Schutztruppe.

179

officer there, General von Lettow-Vorbeck, who inspired both the Germans of the *Schutztruppe* and their native *Askaris*, soldiers. Though their spirit was strong, their artillery was weak, consisting of just the one battery of six 9cm C/73/91 in 1914.

These guns were augmented in April 1915 by some 3.7cm cannon (probably Maxim automatics, although possibly the old Gruson revolving guns) smuggled in by a German steamer slipping through the blockade of the British navy. More guns were supplied less willingly by the *Kleine Kreuzer*, little cruiser, *Königsberg*. This ship was blocked in the delta of the Rufiji River by a superior British fleet and then sunk by its crew. The ship guns had been transported to land first. Their heaviest calibre was 10.5cm (4.2in), very much like the same calibre *Schirmlafetten*, fortress guns. They and the other guns, 3.7cm (1.5in) Maxim Pom-Poms, were then transported by road to Dar es Salaam and put on makeshift carriages, while the ammunition, saved from the sunken ship, dried in the shade.

A real *Ostafrikanische Artillerie*, East African artillery, only existed after another blockade runner came through in March 1916, bringing two 7.5cm (3in) mountain howitzers with 1,000 cartridges for each, and four 10.5cm (4.2in) light field howitzers, also with 1,000 rounds each. These guns enabled the troops to field two batteries of FH and one mountain battery. The guns were pulled by eight or ten oxen each, or in heavy roadless terrain by 40–50 men. The mountain guns were carried by pack animals or long lines of porters. Ammunition should have been a problem, but was not, since a continuous supply of captured British or Portuguese mountain guns also brought their ammunition with them. Thus the last gun of von Lettow-Vorbeck still able to fire when the ceasefire came on 25 November was a Portuguese mountain gun, which he demanded be counted as one of the guns Germany had to surrender.

## THE BATTLES OF VERDUN

There was indeed more than one battle of Verdun – two, to be exact. The first was in connection with the German rush towards Paris which passed Verdun, concentrating on smashing the barrier forts along the Meuse between Verdun and Toul. The second was the battle which showed artillery at its worst: the great annihilator.

### The Rush through the Barrier Forts in 1914

Verdun is on the old and important road from Metz to Paris, where it crosses the Meuse. The town was fortified long ago, first by Vauban around 1700, and again by Sère de Rivière after 1875. Then a ring of about twenty masonry forts on the hills surrounding Verdun was built to keep enemies away and to protect the town from their artillery. The growing range of guns forced France to add another ring even further away after 1887, this time in concrete and with guns under armour, a *zone principale de defense* of 44km (27.5 miles) in circumference. This also included countless concrete shelters, trenches, ammunition stores and batteries in the intermediate zone between the forts, in which to place the interval troops. One of the decisive measures came in 1889 when sheltered rooms for 5,000 soldiers were dug into the rock on which the citadel had been built.

From this cornerstone of French defence to the next fortified city, Toul, a 60km (40 mile) chain of barrier forts covered the Meuse along both sides. These were the targets in 1914 when the German thrust aimed directly at Paris. During the course of this, Verdun itself was left alone; only the right German flank was guarded against an attack out of the fortress town.

The attack started in the classic way, with big-calibre guns shelling the huge barrier forts along the river. These were garrisoned by about 1,000 soldiers each, but were not all of the most modern type. In the course of this attack, some, such as Genicourt, Troyon and Camp de Romains, were reduced to ruins by the German artillery. In these cases the German 15cm and 21cm guns and mortars were augmented by Austrian 30.5cm (12in) heavy Skoda mortars. These were transported in separate loads on a road train designed by Ferdinand Porsche, later of Volkswagen fame. A gasoline engine on the front car of the train drove a generator, which fed the electricity into motors sitting in the hubs of all the wheels, on both car and trailers.

The effect of the shelling was similar to that of the barrier fort of Manonviller. The garrison gave up, even before the concrete ceiling or armour cupolas had been penetrated. But staying inside the forts was no longer possible; the soldiers feared they may become crazy because of the explosions, which shook the forts like an earthquake, blinded the soldiers with clouds of dust and smoke and made them suffer from head-, tooth- and earache. Only Fort Camp de Romains had to be stormed by engineers, and the garrison was permitted to withdraw with full military honours, the officers keeping their swords.

Then the *Marnewunder*, the miracle of the Marne, occurred: the German army was stopped by Moltke the younger, who erroneously feared a French thrust between two separated columns, and the war dug in and continued in the trenches. Moltke was replaced unofficially in his authority as Chief of the General Staff by the Prussian Minister of War, General von Falkenhayn, a very energetic man. Nominally Moltke remained chief; the public was to be spared the fear of a setback.

The French supreme command also erred. They took the fall of the Belgian fortresses of Liège, Namur and Antwerpen, which had been hammered by German artillery into a quick surrender, as a sign that the days of the forts were over, and disarmed their own, denuding them of the guns (except for those installed in the armour turrets), which were then distributed to the field army. So the forts were empty of men and weapons after the middle of 1915; on the other hand, lines of strong field fortifications were dug into the ground some 5–6km in front of the ring of the forts.

## The Battle of Verdun, 1916

The German OHL, the supreme command, had been looking for ways to restart the *Bewegungskrieg*, moving war, again; this suited German tactics and traditions better than the present war of positions. Von Falkenhayn combined this with another aim: a war of human attrition. He calculated that by attacking an important object, which the French simply could not afford to lose, they would be forced to send in their best troops and would then lose them. Even if for every Frenchman killed during this, a German was also killed, Germany could afford to pay this price as her population was larger than that of the French, since France had turned into a country of 'two children families'. It is incredible but true that German operative planning was a simple: *You lose one, I lose one. Let us see who wins in the end.*

The object selected for this inhuman barter – of which the troops were of course not informed – was the fortress of Verdun. But it is not known whether it was intended to take it. The aim was to let the *Todesmühle*, the mill of death, as Verdun was later called, to have its fill of humans.

The preparations took only a few weeks, then nineteen German divisions and 1,250 German guns of all calibres were assembled. They were to smash the French infantry and artillery positions, after which the German infantry was to attack. These preparations were also recognized by the French, who evacuated all civilians in this region, except for 4,000 men destined as labourers, who remained protected inside the subterranean citadel.

On 23 February 1916, the first battle of attrition in the world started. It lasted for 300 days and at the end over half a million soldiers had been killed on both sides. Since then the name of Verdun has been associated with hell – the returning German troops said they had been in *der Hölle von Verdun*, the hell of Verdun, seen mass death and the killing power of artillery. For the battle of Verdun was one of artillery.

Still on the east bank of the Meuse river, the German attack ran against the modern French forts of the outer ring. A few of them were captured – the 'coffin lid' Douaumont almost empty, by a combination of surprise and luck; Vaux after fierce fighting with flamethrowers in the posterns of the fort. But the rest held, not only against storming infantry, but against shelling by the famed super-heavy 42cm mortars. What had happened?

The newer French forts had been constructed after 1885, after the arrival of the high-explosive shells, no longer filled with blackpowder but with modern nitro-compounds, starting in Germany with guncotton in 1883, the shells therefore being named C/83, construction of 1883. These had penetrated the top of Vauban's old masonry forts with their hardened pointed noses and detonated inside, blowing the vaults apart. When the crisis of the explosive shell became apparent to the military world, the interim protection was to take away the earth hitherto covering the masonry and put sand in its place, 1–2m (3–6ft) high, covering this in turn with 2–3m (6–9ft) of concrete, still without iron in it. The fuse of the high-explosive shell would thus be set off when it hit the hard concrete, detonating more or less inside this. The sand filling damped the shock of the explosion, distributing its force onto the masonry vaults beneath it. This proved a wonderfully efficient combination, even better than the later protection by concrete alone, especially if the concrete was not armed and put on in separate layers, resulting in a shingle-like structure.

We have seen the wrecking effect of 21cm, 30.5cm and 42cm semiarmour-piercing German shells hitting this Belgian concrete at Liège in 1914. At Verdun this sandwich design, a makeshift solution at the time, worked better than expected by either the French or German military. Hitting the concrete provided the shock necessary to set off the base fuse, which then detonated the shell with a certain delay. In the time between, the shell penetrated into the concrete, but not through it. The reason was simple: penetration in low-angle fire depends mainly on the energy of the shell at this point of its trajectory. In high-angle fire the shell is fired up into the sky with a low mortar charge until its energy has been spent, and then starts falling down again, thus arriving at the target with only the kinetic energy gathered during its fall, which is a lot less than the big charge of a flat-trajectory gun. What had been sufficient for Belgian concrete was no match for French concrete plus sand. The forts were not penetrated from above. But they were from the rear, because they had been intentionally designed this way. After German troops had moved into the casemates of Fort Douaumont, and were feeling safe there, they experienced an unpleasant surprise. French artillery firing from the other bank onto the fort managed to hit and penetrate these casemates, which lay in the gorge on the rear side. This normally safe position had been built with relatively thin rear walls so they could be easily penetrated by the French artillery in case an enemy took it.

In spite of von Falkenhayn's plan of a limited attack, forty-seven German divisions, almost half of the forces on the western front, were hurled into the *Maasmühle*, the Meuse mill, some of them twice. The German troops melted away, the 1st infantry division losing 11,000 of its 18,000 men within three months. The artillery fired only *Ersatzgranaten*, inferior surrogate shells, now (these must have been the *Kanonengranate*, *see* page 195); the initiative was no longer with the Germans.

On 11 July a turn to defence was ordered, and on 2 September the Battle of Verdun was stopped. The German losses amounted to 328,500 men, while the French had 348,300. By 28 August von Falkenhayn had been replaced by the supreme commander in the east, *Generalfeldmarschall* von Hindenburg, of Tannenberg fame.

Now some details about the German artillery at Verdun. When the attack began, the OHL had not been able to provide all the heavy mortars necessary. Thus these were replaced by *Minenwerfer*, mine launchers, firing against the front lines: twenty-two heavy, seventy-four medium and fifty-six light MW. Altogether, the three corps attacking had thirty-two heavy, eighty-eight medium and eighty-two light MW, for which until February 9,120 heavy, 28,500 medium and 69,600 light *Minen*, mines, in reality shells, had been provided. Fire was to be opened on 12 February 1916 at 08.00; for the heavy artillery from 09.00 onwards; while the MW were to open fire at 13.00. The closest targets were not to be fired upon after 17.00. When the infantry had taken a certain height, the batteries were to change positions. This had to be prepared a long time in advance and be done in sections, so that a strong fire was maintained. The supply of ammunition for this had been cut down by the OHL following requests from the *Armeeoberkommando* 5, the command of the 5th Army, but was still impressive. To every gun was allotted per day: field guns – 300 rounds; light field howitzers – 400 rounds; heavy field howitzers – 180 rounds; mortars – 120 rounds; heavy flat-trajectory fire – 100–250 rounds depending on calibre and rate of fire; super-heavy high-angle fire – 50–100 rounds. (Later, lack of artillery ammunition forced the OHL to gradually reduce these numbers: for FK and lFH the totals dropped to 100 daily rounds in May, to 60 by the end of June, and on 15 July to only 15.) This amounted to 187,320 rounds of all calibres for one day only, six of these *Tagesraten*, daily rates, amounted to 1,124,000 rounds.

Three daily rates were to be transported into the batteries. For this purpose 213 ammunition trains brought their lethal freight from 12 January onwards. The roads were not all good enough for truck transport, especially at this time of the year, so narrow gauge tracks were built between the ammunition parks, the large depots in the rear, and the batteries. When these had been hit and could not be used any longer, horse carts had to take over. Sometimes, because of the terrain, it took up to ten horses to pull an ammunition wagon with fourteen rounds! Other regiments transported their ammunition a different way: they loaded it onto makeshift packsaddles, with two to four rounds per horse.

Due to the soft wet ground, the shells, transported only at night for secrecy, also sank into the ground after unloading, despite the fascines and grids underneath them. The ammunition had to be protected from the moisture, although the men were not, despite complaints that 'their boots were rotting on their feet', as a battery commander complained. This soft ground with its cratered surface also made changing emplacements a chore for both men and animals. For example, when the 3rd battery of *Fußartillerie* Regiment 3 had to do this on 27 February, it took over nine hours to cover 2.5km (1.6 miles) as the crow flies. Each gun had to be pulled by up to twenty-four horses and forty to sixty men on the ropes. It was winter – what a wonderful time to make the soldiers sleep on the open field, frozen at night; the men must have loved Falkenhayn.

The gigantic number of shells and the reports of soldiers surviving the hell of Verdun gives the impression that every square metre of ground was turned up again and again. It was not. Artillery fire on both sides concentrated on important areas: infantry positions, roads for supplies and reinforcements, batteries, and so on. Between these areas there were more peaceful zones, which were neither suited nor used for the aforementioned purposes.

During the Battle of Verdun, the firepower was equivalent to one 21cm shell impacting every minute, or even every 35 seconds, within an area of one hectare, which is very roughly the same as a soccer field. Now, if you were the infantry lying on a soccer field, where every full or half minute a 21cm shell detonates, you would certainly feel as though you were under heavy fire. This level of fire was kept up by both sides during the full nine months of the battle. It has been calculated by the Ministry of War that between February and August 1916 the heavy artillery fired almost 6 million rounds and the field artillery over 8 million. The same numbers at least were also fired by the French artillery. These shells hit and destroyed a relatively limited area, since there were no great movements, advances or retreats during this time. This explains why the whole battlefield was devastated to such an extent. The ground was poisoned by the explosive and of course also by the gas later used on both sides, to such a degree that it was impossible for plants to grow on it afterwards. It took special hardy pine trees to form a first generation of humus soil again, whereon other things grew. I remember my first visit there in the 1950s, half a century after the battle, and I wondered at hearing no birds in the woods around the area. The still-poisoned earth seemed not to have supported a rich fauna of vital micro-organisms at that time.

Gas was fired against enemy batteries to surprise as many gunners as possible before they put on their masks. Then the gas would only hamper their speed of fire. The shells were marked with crosses, the colour indicating the type of gas. Thus the *Grünkreuz* shells fired on 7 May for the first time against enemy batteries were filled with a *Lungenkampfstoff*, lung poison, such as Phosgen. For delivering gas to the enemy, larger calibres than the 7.7cm FK were preferred. The targets for gas had to be away from the attacking troops, so that these were not also endangered. High ground was preferred as a target as the gas then flowed downhill. Sunlight made the gas disperse, so it was fired at night.

On 22 June gas firing started at 22.00 and went on until 05.00 on 23 June. On that occasion the gas was a success for the attacking German troops, eliminating the French barrier batteries in the rear. Only those in front were still active, as they had only been shelled with *Reizstoff*, a sort of tear gas. On the French side 1,600 men had been poisoned by the gas, and ninety killed. It was not only the soldiers that had to wear masks in these cases; the *Kamerad Pferd*, comrade horse, also wore a big one over his head. On this day alone twenty-three batteries of FK fired 13,800 gas shells. This rate of continuous fire took its toll on the guns. The oil of the hydraulic recoil brakes heated and thinned, prolonging recoil, and the springs of the mechanical recuperators broke. Nowadays a rich harvest is reaped: all sorts of shell fragments and duds are found, some of them still live.

The really heavy artillery had been kept to the rear by the AOK 5, which kept them safe from most French guns, but also forced them to fire over long distances, resulting in greater dispersion. The 38cm guns, for example, were 20km away from Verdun. The long distances mean that observation was difficult, and there was no radio communication then. The infantry feared – not without cause, as they had bitter experience – having to serve as unwilling targets for their own artillery. The trouble was that shifting the fire ahead was done by the watch after a certain time. But the infantry could not always be close to the fire. Thus it happened that fire was no longer on the enemy positions when the infantry reached them; it was either too far ahead to support them, or too close to let them advance or, at worst, it hit the infantry. It was not necessarily at Verdun that a saying still alive today was born: that 'the artillery knows neither friend nor foe, only rewarding targets'.

With the artillery observers forced to keep to positions where there were telephone connections, the infantry had to help itself. It held up large wooden frames, showing the artillery a red/yellow flag (and the enemy one in field grey, the colour of the German field uniform since 1912). This did not work in the woods, however, so the troops showed their presence there by red balloons 1m (3ft) in diameter. They were filled with gas from steel bottles that had been dragged along, and rose over the trees.

Balloons were also used by the artillery for the observers, especially when the morning fog rose from the River Meuse. These were captive balloons, a welcome prey to enemy fighter planes. Other observers were luckier. After Fort Vaux was taken by the Germans, it too was used like the Douaumont, both as a sort of safe shelter for the troops – but only when you had made it into the fort, with most of the soldiers killed when close to the fort's war entrance (there were always two entrances to the French forts, one for peacetime behind a drawbridge over the ditch, and another for wartime, which opened into a tunnel some 100–150 feet before the ditch, leading under the ditch into the fort) by the fire concentrated there – and for artillery observers. The French therefore started digging long underground tunnels underneath the other forts, with shell-proof accommodations therein, and ending a long distance away from the forts in the fields, founding the later Maginot line-style of entrance.

At Fort Vaux the French had blown up the eclipsing 75mm twin turret, a sort of modern small-calibre Galopin (the type used again in the Maginot line later). German engineers now cleared away the concrete pieces and found the armour turret and the guns still intact. So one of the gun tubes was used for transmitting messages by light signals through it. But Vaux was no summer camp, as the Germans had to fight the same enemy as the French defenders before them: thirst. The daily ration was two bottles of mineral water for three men, but only if the bearers made it to the fort alive.

This was not a problem for certain other observers, whose post was right on the water. The canal of the Meuse between Bras and Vacherauville had been no man's land for some weeks, when on 13 September *Fußartillerie* battery 793, armed with 15cm high-angle guns, was ordered to post an observer there, 150m (450ft) behind the French lines. This was done the next day, with the men crossing the canal to the other side, where an old barge was tied. During the following night a telephone wire was pulled to the barge and at 06.00 on the morning of the 16th, the observation post was connected to the staff of 8th company of IR 16. The three observers held this post for a long time, directing the fire undetected for three months. Then a major French attack forced them to retire by swimming back across the canal.

One action that took place during the battle of Verdun, and which has been reported in many different versions, was the capture of the strong fort Douaumont. (The French always named their forts after the locality at which it had been erected, in this case the village of Douaumont. German custom had it that forts were named after famous military persons.) It has also been said that the artillery did not play any part in this. This is slander; in one version of the story the artillery was instrumental in the fort being taken.

This is how it happened on 25 February 1916. The 12th infantry brigade had been ordered to 'proceed in the direction of Fort Douaumont' and field artillery regiment 39 to support this. The artillery opened fire at 10.00, but the infantry waited for the order to attack in vain; the telephone wires were constantly interrupted by enemy fire. Finally, at 16.05 the first company started, after an artillery observer had assured them that the artillery was informed of this. Fire from their own heavy artillery impeded the infantry. At 16.30 they were very close to the fort. The signal people of IR 24 had built lines forward, as far as the wire went. The commander of IR 24 was informed by messenger and wire in his command post that 'First lines have reached fort, but is suffering heavily from own artillery fire'. The regiment reported to the rear, but could not get the fire shifted forward, even when other artillery observers reported that men in field grey were to be observed on the glacis of the fort. The staffs did not believe this, since the French gun in the south-east armour cupola was still firing. (This was the 155mm (6in) Galopin-turret model 1896, which eclipsed between shots and rose again for firing, lifted by a clever system of three counterweights. The turret is still there, and next to the iron staircase leading up to it lies a spare tube for the gun: 155mm court, short, with the unique Canet-type breech, taking up recoil by concentric grooves on both sides of the tilting breechblock, easier to examine on the two 155mm court in front of the entrance to the underground citadel.)

The troops next to the fort took cover in the shellholes, firing green starshells from their flare guns, a sign for the artillery to shift fire forward. But these could not be seen by the artillery in daylight. Then *Musketier*, rifleman, Kühn climbed the fort and waved an artillery flag from the highest point to stop the fire – still in vain. In the end it seems that the troop looked for shelter inside the fort, capturing it this way.

The official version is almost as funny. It starts with a German patrol of a dozen men reconnoitring the neighbourhood first, then finding an opening in a caponnier to the inside of the fort, thereby greatly surprising the remains of the former garrison, which since summer 1915 had been reduced to about a dozen soldiers. The French surrendered. The clever commander of the victor's company, 8th company of infantry regiment 24, a First Lieutenant Brandis, who had not been with the patrol, then reported this at once to the staffs and in the end received the highest German decoration, the *Pour le Merit*, and the fame of having conquered the strongest fort of Verdun.

Now for some justice for the poor forward observer, blamed by the infantry for their losses, either because of insufficient support or even their own fire. The forward observers of the artillery went into the trenches with the infantry. But these were not an ideal observation post. It was not possible to look around for any length of time without getting hit in the head (the steel helmet did not stop a rifle bullet). And wires were broken more often than they were working. To direct the fire onto an enemy target, the observer must see this exactly. The probability of a hit is always low. In the case of a battery of 7.7cm FK 96 n/A firing at a target 5m (15ft) deep at a range of 4,000m (12,000ft), the probability of hitting this is only 9.7 per cent. If the observer makes a mistake of only 25m (75ft), too much or too little, this already low probability will sink to 4.6 per cent. If their own infantry and the observer are 100m (300ft) away from this target, dispersion will make the short rounds drop 44m (132ft) short of their own troops with a 25m margin of error, or even as little as 19m (57ft). This then was the cause of the infantry's complaints about their own artillery firing at them. It was also the reason why, in cases where their own troops were too close to an object, the artillery could not fire at this for fear of hitting them.

And then there was dispersion. In the case of the 21cm mortar battery firing with charge 5 at 5,750m, dispersion meant the short rounds dropped 74m (224ft) short of the target, without any observer error. To these 74m would have to be added the range of fragments of an exploding 21cm shell.

A few more highlights. One nice trick was used by German infantry when trying to take the *Haute Batterie*, high battery, a strong French position and 'a thorn in the flesh of 50th ID', by storming it. This action was supported on 2 July 1916 from 02.00 onwards by two heavy MW, firing at regular intervals, with the infantry waiting in the German trench opposite as their whistling approach forced the French defenders to take cover. It sounded something like this: plop, plop (the MW firing) … kawoom, kawoom (two lots of 50kg (110lb) explosive). When the infantry was really close, after two more live mines kawooming, the next pair was fired without fuses: plop, plop. Right after the double plop, the Germans jumped up and were inside the position, while the defenders still had their heads down, waiting for the explosions. This secret may be revealed now – and it only worked once, anyway.

Another foul trick was used when taking the village of Douaumont. The French had fortified this with concrete shelters and fielded a lot of machine guns there, almost impossible to hit with artillery fire of the smaller calibre available. During the preparatory firing, the German artillery shifted their fire forward at 10.45 and some German infantry men jumped up out of the trenches. The French quickly filled their trench to repulse the expected attack. But it had been a ruse: German fire returned, causing losses to the French trench. This was repeated at 11.45, with the same success. At 13.15 the fire shifted again, but this time the infantry jumped out of the German trench and was into the French one before the slow-to-react defenders recognized that this time it was no ruse.

The town of Verdun was also shelled. Guns of 10cm and 15cm and one 38cm gun tried to hit the railway station and the bridges over the Meuse, to interrupt French supply lines. They did not hit these targets, but managed to reduce the town itself to rubble. The same failure occurred with the three 38cm guns firing at the only connection that the almost completely surrounded Verdun held to the hinterland. This road, later called *Voie sacrée*, sacred road, could not be interrupted permanently despite no less than seven dirigibles guiding the fire. The 3,500 trucks bringing supplies to Marschall Petain's troops were not stopped.

In summer 1940 the German *Wehrmacht* returned to Verdun. Within a few days all the forts, some of them modernized, were taken, almost without loss on either side.

The town of Verdun was rebuilt after World War One. The forests around it are green again. Well-paved roads lead to the places of former slaughter and a sign of memory, the fortress-like church of the memorial. There and on many places around Verdun one can see the large fields of crosses, white for the French dead, black for the German. The entrances to these military graveyards are guarded by the same guns that killed their inhabitants. The name of Verdun has remained a symbol for mass death, for the senselessness of war itself. All politicians eager for a war should go to Verdun first for a tour of Douaumont, Vaux and the memorial building, filled in the lower rooms with pyramids of bones and bone fragments.

# 5 Ammunition

You can recognize an old member of the artillery by two means: one is by the hand held behind the ear when you speak to him; the other is by the way he turns red when you tell him that 'the effect of artillery is delivered not by the gun, but by the ammunition', the gun being only a platform, a launcher, which allows the ammunition to do its work. This wisdom is neither new nor cheerfully accepted by gunners, but it is true nevertheless. It could be proved more easily in the old days of blackpowder, which served both as a propellant and an explosive shell filling and could also work without a gun, but is harder with the smokeless powder of today, more accurately described as *rauchschwach*, weak in smoke, in the German custom.

Any advance in gun design was inevitably followed by another in the ammunition field. But this was recognized by only a few. Even if guns did not provide the full splendour that a fiery horse gave to the cavalry, at least they could be shown around and were in most cases of impressive sight and weight. With ammunition, all you could do in the days of evil-tempered blackpowder was keep it dry and hide it from the fools wondering at the black sand while smoking. Thus ammunition remained in the background, perhaps because it was still associated with black magic and the devil. To this day you can take a book on weaponry, open the chapter on firearms – whether hand guns or artillery – find your gun, and not a single word is mentioned about the thing that makes the arm work: ammunition. And the ammunition of today is really just a black box, something the soldier receives well packed and only unwraps to shove it quickly into another black hole, where it soon disappears with a big bang, a flash and some smoke. The ammunition specialist of the army, navy or air force is not expected, nor even permitted, to change anything on the ammunition. So only a select few know their way around it, while all the others smile peacefully, thinking of the old proverb: 'Where ignorance is bliss, it is folly to be wise.' Your bliss is going to end right now.

## THE DEVELOPMENT OF AMMUNITION UNTIL WORLD WAR ONE

From the days of the changeover from the smoothbore muzzle-loader to the rifled breech-loader, introduced by the Prussian field artillery in 1859, the development of ammunition resulted in a completely new generation: the spin-stabilized long shell. This replaced the old sphere-shaped iron ball and had two great advantages over this: one was that its L/, the ratio of length to calibre, was greater than the '1' of the ball, being in the range of 2.5–3, and increasing later. This gave the shell a large cross-sectional density, meaning simply that a lot more iron was pushing behind the point in ratio to the calibre. This resulted in less drag in flight and deeper penetration into the target. The other advantage was that now the same part was always in front, making a working impact fuse possible for the first time.

Many other changes were made to the shell over the next fifty years: spin transferred by soft lead coat, later harder copper rings; the shell made of cast iron, later pressed from more solid steel; and the shell filling changing from blackpowder to high explosives. In addition, sooty blackpowder (the

*Ammunition*

white smoke is the potassium oxide, generated from the combustion of the saltpeter) was exchanged for grey smokeless powder (its black smoke is caused by the carbon atoms left unburnt, because there is not enough oxygen in the powder molecules to go around, just as with an old diesel engine). The tubes grew longer as the new powder pushed slower, and were made of better steel too (as pressure increased with the amount of powder necessary for longer ranges). Then the guns themselves no longer recoiled, but left this to the tube alone, although this was of no importance to the ammunition.

This was the situation regarding German ammunition on the eve of World War One.

## LOADING THE AMMUNITION

Ammunition was placed in the front or rear of the tube, making the gun a muzzle- or a breech-loader. Of the first type, the German army had one gun, the *Minenwerfer*, in 25cm, 17cm and 7.7cm (10in, 6.8in and 3.1in) calibre. All other artillery weapons were loaded from the breech end.

The full round was assembled inside the gun in one of the following ways.

### Separate

The ammunition was loaded in three stages, and only with breech-loading guns, starting with the shell being rammed home, mostly with the tube in the horizontal loading position. Then the propellant was shoved in, in one or – with larger calibres – two to three charges for guns, and even more for howitzers, the last one always inside a metal cartridge case. This was needed for obturating the modern Krupp *Keilverschluss*, the wedge-breech, and also contained in its centre the primer screw, which transformed the kinetic energy of the hammer pulled by the lanyard in firing, into physical energy of heat; the flame then ignited the powder. Breaking up the charge into portions was necessary for handling its weight in big cannons; the weight of the charge was not normally changed, but it did save wear and tear on the tubes in the case of howitzers and mortars. With these, only the amount of powder necessary to fire at the desired range was loaded, so tube life was prolonged.

The bags of modern nitro powder were – and even today are – kicked into deflagration by a tiny charge of blackpowder first. Before the metal cartridge case appeared around 1890, obturation worked by a metal ring-system, of which the Broadwell obturating ring was the best known. But metal cases already containing the primer gave a higher rate of fire, hence the guns using them were called 'quick-fire guns'.

Not all artillery guns in Germany used this wedge-breech. The muzzle-loaders with their permanently closed breech end obviously did not; neither did one or two of the mortars.

We now need to take a slight departure from the discussion of ammunition. Artillery had developed a three-class system over the centuries: the guns or cannon, firing in the low-angle mode with elevations from 0 to 45 degrees; the mortars firing in the high-angle mode with trajectories between 45 and 85 degrees; and the howitzers, able to fire in both. They could also be recognized easily by the length of their tubes, if seen, or expressed by the calibre length, L/. Cannons had the longest tubes, their L/ at 30–50 by the time of World War One. Then came howitzers with an L/15–25, followed by the mortars with their small L/8–12. In certain countries, however, some mortars were made into howitzers. And today, after the interim use of the name 'cannon-howitzer', you can no longer depend on this term, as can be seen in the case of the trilateral development FH 70 (in Germany FH 155-1), which is called a howitzer in spite of an L/39, which would previously have made it a cannon. In World War One, German artillery, from 105mm calibre (the light field howitzer) onwards, loaded separately, using metal cases for obturation.

We return now to the mortars without wedge-breeches, such as the 30.5cm Beta- and the 42cm Gamma-mortars (but not the 42cm M-Gerät, *Dicke Bertha*). They had the screw-type breech, named after deBange, though he had not invented it. This

188

breech was obturated in most models by a pair of Broadwell rings, though not in these two models; in these a metal case served.

## Semifixed

Loading ammunition this way is done by selecting the correct charge, by removing unwanted charges from the metal case (which arrived filled with the highest number of charges in their bags, the bags not needed) and then sticking the shell on top of the powder into the space left in the shell. To my knowledge, this was practised only on the 105mm US field howitzer, despite its obvious advantages, as it combined the thriftiness of separate loading with the speed of the following type: the cartridge.

## Fixed Ammunition

Fixed ammunition in the form of the cartridge contained within its metal (in peacetime this would be brass, but in World War One Germany tried all sorts of other metals, such as aluminium, iron, and so on, because of a shortage of copper. Today, with the infantry's ammunition-consuming full automatic rifles, machine guns and 20mm automatic cannon, steel is good enough for the cartridge cases) the shell, the – unvariable – powder charge and the primer. It was thus loaded in a single motion resulting in a high rate of fire, always desirable in firing. But since it was also a waste of powder and caused extra wear on the tube, cartridges were only used with guns that depended on a high rate of fire for efficiency. Antitank and antiaircraft guns belonged to this class of larger calibres too, but in most cases the cartridge was limited to small calibres.

A sort of compromise is made by some modern tank guns, which like the 120mm smoothbore gun of Leopard 2 have a combustible case for quick loading without worrying about fuming cartridge cases (CO and NO gases) of fired rounds inside the fighting compartment afterwards, and for obturation bear an extra short steel case at the bottom. But then one can also try to do this with the Broadwell rings, as in the case of the 120mm gun of the British Chieftain tank.

The smaller German calibres such as the 2cm and 3.7cm automatic guns, the 3.7cm and 5cm guns, the 5.7cm and the old 7.7cm FK 96, and the 10cm (105mm) guns, all fired cartridges.

The increase in rate of fire is best shown using the different German fortress guns in the same armour cupolas: the 15cm howitzer of 1893 with separate loading fired two rounds a minute, the 10cm cannon of 1900 fired nine cartridges in a minute. But, due to its choice of weaker charges, the howitzer could fire with a curved trajectory, reaching deep into the ravines of hilly ground. The flat trajectory cannon fired straight over this dead ground.

The German gun designers of Krupp, the military of the APK and the gunners had all opted for the metal case for their guns before World War One, fired out of artillery pieces with the wedge-type breechblock. This choice had been motivated by the desire for a high rate of fire. No one had imagined the tremendous rise in ammunition consumption to be experienced in World War One, which caused terrible problems for industry in providing the necessary fuel for the war machine. The signs had been there since the Russian–Japanese war, but again no one had read them correctly.

# PROJECTILES OF GERMAN ARTILLERY

The components of the ammunition were certainly not much different to those used by other European countries, except for the names of the chemicals involved.

## Shells and Other Projectiles

The artillery of all countries relied on three different types of projectile for centuries. The first was the solid ball, which later became the armour-piercing inert (without explosive) shot. The next was the explosive bomb of the mortar, which was then used by all guns as the explosive shell. And

## Ammunition

the last was the multiprojectile round, either canister or shrapnel.

The German artillery went into the war with shells and shrapnel as the main projectiles, and a few special ones, such as canister for close defence and starlight shells for illumination. A typical supply was that for the 10cm gun in the forts: 50 canister-, 2,600 shrapnel-, and only 350 shell cartridges. The appearance of the British tanks then put extra emphasis on armour-piercing shells.

Shortly before, in 1907, Germany had erred by pinning its hopes on a single projectile, combining both the effects of shrapnel and shell, and therefore called *Einheitsgeschoss*, universal projectile, easily switched by hand from one function to the other. This had been made possible through the idea of Dutch First Lieutenant Pieter Daniel van Essen to fill the shrapnel with the explosive of the shell between the lead balls, where hitherto only a filling producing smoke had been in order to show where the shrapnel had been on its trajectory at the moment it sent out these balls. This only worked with the new TNT of 1902, as the old picric acid of 1888 would have caused dangerous picrates sensitive to shocks by reaction with the metal balls. The van Essen patents were bought by Ehrhardt of Rheinmetall, who then sold a *Brisanzschrapnell Ehrhardt-van Essen*, a high explosive shrapnel E-vE. (The French followed by also producing something of this nature, the *Obus Robin*, the Robin shell, but did not introduce it. Besides Germany, only Austria adopted this combination projectile.)

Of course, this multifunction also required a special fuse with no less than four different functions. To start with it was a dual fuse, working either on impact as a shell or as a time fuse on the course of its trajectory. Then it could be switched to work on impact either as a super-quick fuse or with delay. In the case of the powder train fuse, *Haubitzzünder* 05 (HZ 05), the howitzer fuze of 1905, this took no less than 109 separate brass parts. These tiny bits were produced by Krupp and assembled in the government fireworks laboratories (military ammunition plants) at Spandau and Siegburg.

The *Einheitsgeschoss* was introduced for the FK 96 n/A, besides the shrapnel and the shell, as *Einheitsgeschoss* FK-G 11, and for the lFH 98/09 as the only projectile, the *Einheitsgeschoss* FH-G 05. They had all the drawbacks of a compromise, being no better than a shell, but much more expensive in terms of fuse and construction. In light of the great demand that the war placed on the ammunition industry, which this was hard pressed to satisfy, the *Einheitsgeschoss* was soon done away with.

For history's sake, here are the data of that used for the FK 96 n/A: the FK-G 11. Its powder train fuse was without an alternate delay function, but almost as complicated as the HZ 05, only it was now made of aluminium. The 7.7cm projectile weighed 6.85kg (14.4lb) and contained 294 balls of hardened lead at 10g (0.3oz) each. The explosive filling was 0.25kg (0.525lb) *Füllpulver* 02, TNT, and the lead balls were fired in the shrapnel mode out of the base by 80g (2.8oz) of rifle powder 71, blackpowder grains.

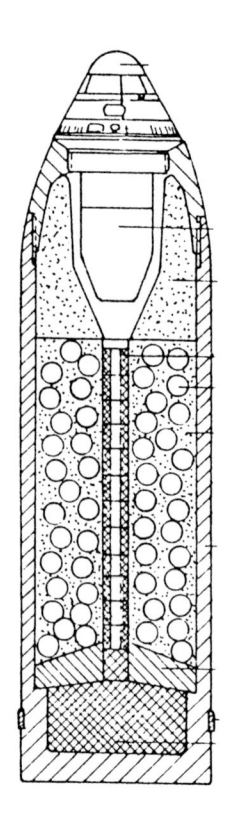

*The erroneous way of the* Einheitsgeschoss, *the universal shell, supposed to work both as a shell and as shrapnel, was ingrained in German artillery from 1905 onwards, until the war showed this up as a mistake. Seen here is the 105mm* Feldhaubitzgeschoss 05.

## Shrapnel

German shrapnel such as the 7.7cm *Feldschrapnell* 96 for the FK 96 were of the so-called *Bodenkammer*-type, base chamber-type, where the blackpowder charge lay at the base of the projectile behind the balls. This was ignited by the fuse after the set time had run out, through a hollow tube running from the fuse to the charge. The gases of the powder tore away the purposely weak connection between fuse and body of the shrapnel, and at the same time pushed a steel disk forward and by this the balls, which were fired out of the front of the shrapnel. This resulted in them reaching a higher speed, and also aiming at the target, unlike the earlier shrapnel types with *Kopf*-, head-, or *Mittel-Kammer*, middle chamber, where the charge was at the indicated place, making the shrapnel eject the balls to the rear with less efficiency. The balls were fixed by a mixture of phosphorus, colophonium and blackpowder in the space between them, which now also burned, giving off a dense black cloud (caused by the colophonium, not the blackpowder), to show where it had exploded, either before or over the target.

The shrapnel contained the same number of balls, 300 at 10g each, as the *Einheits-G.* did. Due to its thinner steel walls of 5mm (0.2in), it was also of the same weight as the shell, in spite of all the lead balls inside, enabling the gunners to fire it with the same firing table as the shell. For the super-heavy flat-trajectory guns of 38cm firing at the French fortifications, shrapnel to combat troops was made later. The long time of flight also meant that another mechanical fuse with a longer running time was necessary.

After the war, long discussions in military circles about the role of shrapnel in future wars resulted in its abolition in the mid-1920s.

## Canister

Canister was used for close defence work only, dating back to the days when cavalry would attack a battery, preferably when this was on the march. In this case German artillery used to hold two rounds of canister ready on the gun in metal boxes. Most guns received canister in the days before the machine gun, but it has since returned to favour, even after World War Two. I would certainly not like to be in the way of a beehive round, when its 4,000–6,000 (according to calibre) *flechettes*, tiny steel arrows, swarm out.

A typical German canister of this time was that of the 10cm *Kanone*, which had been designed in the pre-machine-gun era, when canister was still held in high esteem. It consisted of a metal can filled with lead balls fixed in a sulphur casting. The powder gases forced this out of the muzzle, with the spin and air resistance opening it, so the balls spread into a pattern, covering a certain area. The kinetic energy of the balls was deemed sufficient up to 600m for an enemy infantry soldier.

## Shells

German shells used at the beginning of the war were a heavy thick-walled version. For the FK 96 n/A, the 7.7cm *Feldgranate* 96, field shell of 1896, was used. It was of the same weight as the shrapnel – 6.82–6.85kg (14.3–14.4lb), depending on whether the fuse was made of brass or aluminium. It contained only 220g (0.5lb) *Füllpulver* 02, filling powder of 1902, TNT, and the older ones 190g (6.1oz) *Granatfüllung* 88, shell filling of 1888, picric acid, since there was not space inside for more. But the design relied on the thick walls of 18–20mm (0.7–0.8in), resulting in heavy fragments on explosion, and was thus of more effect against 'soft targets', as the poor infantry man is known today. The French shells relied on the blast effect of more explosive detonating in a thinner shell. Each recipient liked the other gift better, resulting in Germany introducing the so-called *Minengranaten*, mine shells, for the artillery during the war. The *Minenwerfer* had used them since before the war.

Special shells such as smoke or illuminating shells were of the same carrier-type as the shrapnel, ejecting their cargo in the air, with one sort of smoke shell also bursting on the ground, combining this, because of their phosphorus filling, with an incendiary effect. This also goes for the gas shells, where normally the filling was contained in

Ammunition

a glass bottle, safe during storage and broken by the shock of firing.

The bodies of the shells had first been made by casting them in iron, but this was no longer shock-proof enough for either the modern explosive fillings after 1883 or the muzzle velocities achieved in 1914. They were now cast from steel, or pressed from this using the method invented by Ehrhardt, director of Rheinmetall. This required a square block of hot steel to be put into a round mould, and the block then perforated with a round male die. This resulted in a steel tubing which could then be machined. It took Ehrhardt some time to perfect his invention, since it would not work with a round block of steel. Ehrhardt later used this invention, which became famous worldwide, as a basis for the trademark of Rheinmetall: a square inside a circle. Since this form of manufacturing shell bodies needed heavy presses, the shortage of shell bodies was programmed in before, when demand rose unexpectedly at the beginning of the war. We shall return to the shells when we look at the developments after 1916.

**Propellant** Propellant was no longer blackpowder. Soon after the new nitro powder had been introduced together with a small 8mm calibre in the form of the *Gewehr* 88, the rifle of 1888, its obvious advantages were also used for larger calibres. But a change like this required a lot of money, and so for a long time the old blackpowder-filled shells of the 9cm C/73 were still fired with blackpowder, until the C/73/88 and C/73/91 changed this step by step. Now charging the gun was done the international way, with the shell being rammed home first. This was followed by the minimum charge of powder necessary for the desired range. The firing table told the gunner which charge and elevation should be used for which range. So for the lFH 98/09 one of the *Ladekanoniere*, loading gunners, took the brass cartridge case and pulled the cardboard cover on top. Inside were the charges – seven altogether for this gun – in inverted order. On the bottom of the case lay the basic charge 1 on the primer screwed into the base of the case. This charge 1 also held an amount of blackpowder on the underside, next to the flash holes of the primer. This was to kick the safe and phlegmatic nitro powder into action. On top of 1 was 2, then 3, and so on, until charge 7 on top.

The officer in the gun direction centre had already found out from his firing table that this mission of fire needed charge 4 to reach the target. So he ordered 'Charge four', whereupon the loading gunner pulled out the number 7 on top, then 6, and so on until 4 could be seen. Having checked this, the cartridge case with the rest of the powder was loaded and the breech closed. The gun was now ready for firing, as the primer was installed in the factory; it only needed a firm tug on the lanyard. (Certain nations used charges filled with bags of different colour. This was so that the illiterate gunners of their colonies could recognize the charges. This system must have been better, since it resulted in victory in two wars.)

**Primers** The primers igniting the powder were of the percussion type. The pull of the lanyard made a sort of lever-like hammer impact on a firing pin, which in turn hit the base of the primer in the cartridge case. This resulted in a flame from the primer, which ignited the powder. The only gun with the old type of friction tube primers was the old C/73, still fielded by reserve troops during the war, and also used as a makeshift antiaircraft gun at first.

**Fuses** Fuses make the explosive inside the shell go off at the desired time. They must be absolutely safe at all other times, during storage, transport, handling, and even on the first part of their trajectory. This was done by inserting parts in the *Zündkette*, the train of ignition of the impact fuse, which had to be removed after firing, before the fuse could detonate the explosive filling. This was achieved in the *Granatzünder* 96, shell fuse of 1896, for the FK 96 and 96 n/A, by a *Pulverkorn*, powder grain, a tiny block of compressed blackpowder (when this is compressed enough, it does not explode, but burns slowly; the igniter cord

192

functions this way, for example) which was pierced by a pin on firing and burned for one second, blocking the firing pin's route to the detonator, thus preventing this being set off. When the powder grain had burned after a few hundred metres of flight, the way was free for the firing pin to act. This happened on impact. Then the pin struck and the detonator cap set off the filling, though not immediately, first detonating a go between, a *Verstärkungsladung*, a booster charge, needed for the really lazy modern explosives. This was with the fuse in 'super-quick' position.

When in the 'delay' position the fuse first set off another tiny charge of compressed blackpowder, which took a few thousandths of a second to burn through, then set off the detonator cap. In the meantime the shell had penetrated into the ground of a field fortification, so it would go off 'confined' and not simply blow its energy into the air. A fully confined shell is equivalent to a shell of double the calibre, or eight times the amount of explosive, turning a 10.5cm into a 21cm and a 21cm into a 42cm.

The shrapnel had a similar fuse, but as the shrapnel did not work on impact, instead distributing its contents in the air, the fuse obviously had to work in different way. The fuse used on shrapnels before World War One was a powder train fuse, in which the shock of firing set a ring of powder burning. This went on until the flame coincided with an opening leading down to the detonator cap. The ring could be turned so that this happened sooner or later. This was a *Zeitzünder*, a time fuse, in the form of a *Brennzünder*, a powder train fuse.

There were drawbacks to this. The flame could be extinguished by the air during flight, or the fuse set incorrectly for too long. In both cases a live shrapnel would land as a dud. This prompted the development of a fuse that would work, if not in flight, at least on impact, by combining the mechanisms of both types. It was called a *Doppelzünder*, double fuse, and artillery often uses it today (if not the more expensive electronic fuses).

This double fuse could be built in the powder train model, but still held the same drawbacks. So a new type of fuse appeared, both as a simple time

*Switching the production of ammunition from low peacetime output to wartime mass production resulted in problems, especially when the trained artificers and experienced workers had to depart to the front-lines and women replaced them. This led to shells detonating on firing or still inside the tube, and killed many a gunner. The problem is shown here on a 21cm mortar.*

fuse and a refined double fuse: the mechanical fuse. This was clockwork, and was already wound in the factory. Firing set it running and at the pre-set time it set off a needle for the detonator. With this sort of mechanical fuse it no longer seemed state of the art to work with the powder grain safety, so the fuses were made safe by in-built mechanisms, which either blocked the firing pin, or – more elegant – made the detonator swing out of harm's way, until the rotation of the shell caused by the spin of the rifling, or the inertia during the firing shock, or both, armed the fuse.

*Ammunition*

And when the antiaircraft gunners complained about their own duds dropping down from the sky onto their own heads, their shells received a *Zerleger*, dismantler, a self-destroying device. This either worked from within the tracer, by using its flame to set off the shell, or in the case of mechanical fuses it was done by clockwork.

All shells were supplied and stored without fuses, these being screwed in only a short time before firing (as it is still done today). The only exception were the cartridges, already supplied complete.

## GERMAN AMMUNITION STOCKS IN 1914

The German ammunition was well designed and carefully made, but had one great disadvantage: there was not enough of it. When the war started in 1914, Germany had stored for her artillery:

- 7.7cm *Feldkanone* 96 n/A
  - for 3,042 mobile guns, 987 rounds for each
  - for 108 fortress guns, 707 rounds for each
  - for 104 others, 457 rounds for each
- 10.5cm *leichte Feldhaubitze*
  - for 720 mobile howitzers, 973 rounds for each
  - for 28 in fortresses, 636 rounds for each
  - for 44 others, 386 rounds for each
- 10cm *Kanone*
  - for 16 mobile guns, 1,425 for each
  - for 176 other guns, 1,425 for each
  - for 152 other guns, 1,000 for each
- 15cm *schwere Feldhaubitzen*
  - for 408 guns of the field army, 1,095 for each
  - for 650 guns of the siege train, 723 for each
  - for 650 guns of the defence artillery, 823 for each
- 21cm *Mörser*
  - 112 mobile mortars, 782 for each
  - 80 others, 500 for each
  - 32 others, 562 for each
- 13.5cm *Kanone*
  - 32, 1,261 rounds

The military considered this to be sufficient. In the last battles fought by the Germans in 1870–71 at Vionville, Gravelotte and Sedan, the guns had fired 88, 55 and 57 rounds respectively per day. The numbers fired during the Russian–Japanese war were only published by the Russians. During the battles of Tashitshao, Liaojang, Shaho and Mukden they amounted to 522, 422, 364 and 504 rounds per day per gun. But all major European powers had assumed that a European war would take only three months, even if it would consume ammunition at the rate shown in the Far East. No one could imagine anything like the major battles of the Russian–Japanese war lasting for a long time. But they did.

This lack of enough ammunition for the artillery to support the attacking infantry effectively, together with the inability to combat covered artillery, was a key factor in making World War One a long enterprise with high losses.

## GERMAN AMMUNITION DURING THE WAR

Hurried wartime manufacture is never up to the quality of unhurried peacetime production. This also applies to ammunition. In the case of the real heavy calibres, such as the 42cm Gamma-mortars and *M-Geräte*, Germany lost a lot of these pieces due to shells exploding inside the bore – premature bursts. These were caused by the explosive filling poured into contained cavities that collapsed during the shock of the firing with sufficient energy to detonate the filling. Other cavities were in the steel of the shell bodies, and were pierced by the gases of the powder, also resulting in premature bursts. In all these cases the tubes split and the crews were injured or even killed. (The days when German gun tubes had to be *sprengsicher*, safe against shells detonating inside them, had gone. Longer tubes with their higher weight no longer permitted the luxury of the additional weight.)

The lack of sufficient ammunition stores soon made itself felt. But more could not be produced rapidly of the usual quality; the heavy presses

*Even the 42cm* Gamma-Mörser *was not spared from shells detonating on firing, in spite of all its armour.*

necessary for the Rheinmetall-process engineering, and even steel, were lacking.

In August 1914 a new type of shell was introduced: the 7.7cm *Kanonengranate* 14, the gun shell of 1914. Its development by the APK had been perfected by the spring of 1914, just before the war began. It was a step back in every way. The shell body was cast in iron and had therefore to be thick enough to take the shock of firing and to produce large enough fragments on detonation. This reduced the interior volume so that the shell only contained 180g (6oz) of explosive, less than the field shell 96 and especially the *Einheitsgeschoss* FK-G 11. And the explosive was no longer the powerful TNT, but at first a replacement of ammonium-nitrate-carbon explosives, such as Donarit, Roburit or Westfalit. They were too sensitive against the shock of firing and caused so many bore prematures, however, that at the beginning of 1915 all these shells had to be withdrawn from the front.

A new filling of *Füllpulver* 60/40, filling powder, a mixture of molten ammonium-nitrate and TNT, later solved the problem. The same sort of *Ersatz* reached the 10.5cm lFH in the form of the *Haubitzengranate* 14, howitzer shell of 1914, another cast-iron shell, containing only 0.335kg (0.7lb) of explosive, a quarter of that inside the FH-shell 05.

The blame for most prematures was laid at the door of the new fuse. This *Kanonenzünder* 14, gun fuse of 1914, had been introduced together with the same for the howitzer (HZ 14) in 1914, to enable the industry to mass produce fuses, as the old method of assembling the countless parts in two laboratories was no longer sufficient. The new fuses were simple impact fuses. But even their production was accompanied by mistakes made by the new manufacturers now enlisted. Cast-iron shell bodies cracked at firing, so the flame of the propellant touched off the *Ersatz* explosive resulting in more prematures. The rate in 1915 was one for every 5,000–6,000 rounds fired, resulting in gun-crews being wounded or even killed.

From 1915 onwards the shell bodies were cast in steel, or even made by machining round bar-steel. These thinner-walled bodies also allowed for more explosive, so that French superiority in this field was compensated. From 1917 onwards enough heavy presses had been installed to return to pressing thin steel bodies for the shells. The rate of manufacture was also raised enormously, from 343,000 rounds a month at the beginning of the war to 11 million at the end.

The development of flash ranging made night firing dangerous. Therefore special additives of antiflash salt (potassium chloride) were loaded in front of the charge at night. With brass becoming rarer all the time, cases and cartridge cases had to be made from steel. These could no longer be made the traditional way invented by Polte at Magdeburg in the 1890s, the 'ball rolling', since this only worked with the softer brass. The steel cases had to be made in two parts, with walls and base, then welded together. This led to cartridge cases jamming or splitting. The troops had to live with this, since making steel cases was only perfected after the war.

## NEW DEVELOPMENTS OF SHELLS AND FUSES

New developments continued to be made during the war. The war of positions from 1915 onwards would not have been possible without artillery; even then infantry could only advance as far as their own artillery ranged. Bringing the guns forward in a terrain full of shell craters was a slow process with many losses, as shown at Verdun. The enemy attacks were stopped by German *Auffangstellungen*, parrying positions, prepared before out of range of the enemy artillery. Also, more importance was laid on stopping enemy supplies and hitting their command posts. This meant extending the range of artillery fire.

For German artillery this began in February 1917 with the 7.7cm *C-Geschoss*, the C-shell, for the FK 16. It owed its increased performance to a shell body designed for less drag, thus preserving its kinetic energy for a longer distance. It too was loaded separately (the FK 16 no longer used the cartridge of the FK 96 n/A) with a special charge. The range was impressive: 10,700m (6.7 miles), but with higher dispersion. This was caused by too little distance between the bourelette and the driving band, which made this projectile yaw in flight.

The next development was that of a sensitive-impact fuse. At the beginning of the war artillery had mainly fired shrapnel against troops in the open. Then shells were fired only in the *Abpraller-Verfahren*, the ricochet mode. In this the shell was fired short of the target on purpose, with such a flat angle of impact that it ricocheted again. The impact fuse was in the delay position, so the fuse, set off on impact, then detonated a short distance further away, hopefully over the heads of the enemy. The terrain, now with its shellholes and trenches, was no longer suited to this method. With output increased in favour of the shells, the effect of their fire was unsatisfying, because the rather slow fuses only detonated when they had already penetrated into the ground, even when fired without delay.

So a new super-quick fuse was needed. One such fuse had already been made before the war by Krupp for firing at balloons. This *Aufschlagzünder für Ballonbeschuss* 09, impact fuse for firing at balloons of 1909, was constructed so that it went off after hitting the thin skin of the balloon. In December 1916 the German artillery introduced such sensitive fuses for FK and lFH, the *empfindlicher*

*Aufschlagzünder* 17, the sensitive impact fuse of 1917. Since there was no more brass available for their manufacture, they were made without it, with the bodies of either zinc or cast iron. These fuses were no longer supplied ready to be fired, the gun crew only having to screw them into the shell. (The nose opening of the shell was and is closed by a plug, which ends in a ring, the *Mundlochschraube*, the screw for the mouth hole, the nose plug or fuse hole plug. If it was forgotten in haste to exchange this for the fuse, you fired the shell with a *Ringzünder*, a ring fuse, alluding to the ring shape of the plug, which of course resulted in a dud.) Now for safety reasons the fuses were supplied without the firing pin, which was inserted at the time of loading the shell. Forgetting this also resulted in a dud, of course. But when (or if) it worked, the sensitive fuse went off on impact, distributing its shell fragments over the surface of the ground, rather than inside it.

## Gas Shells

Gas shells were used by both sides. The mixtures were the same; only the names differed, mostly those of the proud inventors such as Lewisit, Adamsit, Clarksit (all British). The first deployment of gas had already occurred before World War One, and when the war started the gas was let out of the bottle at first, and carried by the wind to the enemy trenches, and then later fired by the artillery. Casualties were high, especially when firing began, when masks did not exist or were ineffective. The shells for this type of warfare were of the same design as the old *Kopfkammer-Schrapnell*, the shrapnel with the burster charge sitting in the nose. This charge no longer ejected a lead ball, but disrupted the thin-walled shell body, so that the liquid filling – mostly contained inside a glass bottle – was set free. The different poisons were either non-resident vaporous gases, effective for a short time only (advisable if you wanted to take a certain area of terrain) or resident sticky liquids, if you wanted the terrain denied to the enemy for a long time. They affected eyes, lungs and skin.

## Armour-Piercing Shells

'Armour-piercing shells' was the cry of the German soldiers when the first British tanks attacked them. A quick solution would have been the reduction of the well-tried armour-piercing shells of the navy. The APK may not have known that there was such a branch of service; in any case they developed their own version of tank killer. This took time, during which the artillery was impotent against tanks, and resulted in a complicated design. Maybe this was forced by problems of manufacture, allotted materials and other considerations. (It is always easy to mock former decisions if their cause is unknown. It should be fair to consider that the old boys had their reasons.) Still, the resulting antiarmour shell *Kanonengranate* 15 *mit Panzerkopf*, gun shell with armour head of 1915, for the 7.7cm FK 96 n/A, was basically the same shell body, with a hardened armour nose in place of the fuse. This was encased behind the nose and detonated the explosive filling of 170g (5.8oz) TNT. The top of the shell showed none of the enhancements for armour-piercing already practised by the German navy, such as an additional 'cap' of soft iron to aid in penetration (it is hard to believe that the armour is penetrated deeper if your projectile has to penetrate extra metal. But the soft iron works in several ways: by protecting the hardened shell point, preventing it glancing off and 'greasing' during armour penetration). And there was no ballistic windshield either, the long sheet-metal cone reducing drag and preserving energy.

The navy had all of these and the army should have copied their design. (On the other hand, the German 3.7cm Pak went into World War Two with a solid armour-piercing round, without cap, shield or explosive: the *Panzergranate*). This shell was again fixed in the cartridge case and filled with a special powder charge for deep penetration, a practice given up with the new FK 16. This was in contrast to the cartridge firing FK 96 n/A which had to be loaded separately. The power of this round was enough for the thin-walled tanks of the first generation.

## Antiaircraft Ammunition

Antiaircraft ammunition was also developed during the war. Many solutions with which to enhance its effect had been tried and refused. The shells of the 9cm C/73/91 or the 8.8cm guns, with their fillings of about 1lb, just did not carry enough explosive. So special shells were filled with pieces of chain, with wire holding barbed hooks, with several smaller explosive charges or with incendiary pellets. These were more effective against balloons and dirigibles, but had little effect against planes. Fuses were improved by exchanging the old powder train types for mechanical fuses, of which two were developed by well-known clockmakers in the Black Forest. One was Diehl using clockwork as a base, whereas Junghans used the centrifugal force of the rotating shell, counting the revolutions. Their improvements could only be supplied to the more modern AA guns of 8cm, 8.8cm and 10.5cm because of insufficient production. Fuses had to be set by hand, slowing down both loading and firing. Particular problems were caused by the lack of copper for the driving bands, which held for all artillery. Other materials tried were zinc, aluminium, brass, cardboard and iron. Zinc and soft electrolytic iron seemed best, but still resulted in too much tube wear for the Flak. In the end a *Kupferpanzerband*, copper armour band, was designed, consisting of iron in the groove of the band and copper on the bearing surface only. This saved half of the copper, but at a cost of problems in welding the copper onto the iron part, only solved in 1917.

Cartridge cases also had to be made of a material other than brass, resulting in the same problems with iron *ersatz*. So for the Flak, only able to combat its targets for a short time anyway, cartridges with brass cases were still supplied.

Powder was another bottleneck. The ammonium powder had already helped a little. The Flak also used special ammunition with a reduced charge when firing *Sperrfeuer*, barrage fire. Since this was aimed into a space, where sooner or later the plane was supposed to arrive, it did not matter if the shell also arrived there sooner or later. This saved both on tube wear and powder consumption. Of course, muzzle velocity also dropped. In the case of 7.62cm antiaircraft guns the reduction was from 590 to 410m/s (1,770 to 1,230ft/s); for the 8.8cm from 785 to 542m/s (2,355 to 1,626ft/s); for the 10.5cm Rheinmetall from 580 to 460m/s (1,740 to 1,380ft/s); and for the 10.5cm Krupp from 720 to 445m/s (2,160 to 1,335ft/s). The altitude reached by the shell fell accordingly.

The tubes of the AA guns also suffered, as firing fixed ammunition (cartridges) they always used the full charge. This consisted of hot nitroglycerine powder, which eroded the lands of the tubes so much that tubes only lasted for 3,000–4,000 rounds. (Field howitzers firing lower separate charges survived twice this number and more.) It was calculated that an artillery piece could throw shells equivalent in weight to the weight of its tube:

- *Minenwerfer*: shells weighing 5,000 times the weight of its tube
- howitzers: shells weighing 500–600 times the weight of their tubes
- field guns: shells weighing 300 times the weight of their tubes
- 42cm mortars and *M-Gerät*: shells weighing 100 times the weight of their tubes
- antiaircraft guns: shells weighing only 20 times the weight of their tubes.

The longer the tube, the shorter its life, was another finding. This was a result of the length of time the hot gases were confined in the longer tube, heating this up during shell travel, until it had left the muzzle and the gases could escape. The Flak suffered the same high number of premature bursts inside the tube as the other artillery. They found the same reasons, and also put it down to the lack of trained and experienced personnel for the manufacture (women had now taken the place of men in the ammunition plants) and final inspection, with the old experts serving in the fields. Sabotage was almost non-existent.

On the whole, production was higher and things were running better at the end of the war.

*Ammunition*

## Shells of the German Light Field Artillery 1914–1918

| Shell Model | Introduced | Shell Body | Wall Thick (in) | Explosive (oz) | Remarks |
|---|---|---|---|---|---|
| *7.7cm FK* | | | | | |
| FK Gr 96 | Before war | Cast steel | 0.56–0.72 | 6.7 | Peace quality |
| FK Gesch 11 | Before war | Pressed steel | 0.19 | 8.9 | Complicated fuse |
| K Gr 14 | August 1914 | Cast iron | 0.81 | 6.4 | Poor quality |
| K Gr 15 | July 1915 | Cast/rolled steel | 0.6 | 13.6 | 8.2oz explosive for a time |
| lg FK Gr | February 1916 | Pressed steel | 0.4 | | Unstable over 5,000 m |
| K Gr 16 | December 1917 | Pressed steel | 0.402 | | Replaced lg FK Gr |
| C-Geschoss | February 1917 | Pressed steel | 0.436 | | Greater range |
| *10.5cm lFH* | | | | | |
| H Geschoss 05 | Before war | Pressed steel | 0.58 | 48.9 | 32.1oz explosive for a time |
| H Gr 14 | August 1914 | Cast iron | 1.22 | 11.9 | Poor quality |
| H Gr 15 | February 1915 | Cast/rolled steel | 0.66 | 50 | |
| lg FH Gr | December 1915 | Pressed steel | 0.55 | 71.4 | 54.3oz explosive for a time |
| C-Geschoss | June 1917 | Pressed steel | 0.7 | 53.6 | Greater range |

*7.7cm FK:* 1. Kanonengranate *16 (shell);* 2. C-Geschoss *(shell with improved ballistics);* 3. Kanonengranate *15 mit Panzerkopf (armour piercing);* 4. Feldschrapnell *96 (shrapnel, filled with steel balls);* 5. Kartätsche *(canister).*
*10.5cm lFH:* 6. lange Feldhaubitz Granate *(shell);* 7. C-Geschoss *(shell with improved ballistics);* 8. Haubitzschrapnell *16 (shrapnel);* 9. Kartätsche *(canister).*

# 6 Tactics, Ranging, Transport and Shelters

*Tactics also changed: no longer did the infantry advance close packed as in this pre-war manoeuvre, with the artillery accompanying it, behind the screen of the cavalry.*

## TACTICS

German field artillery tactics in 1914 still required it to drive up close to the enemy, preferably following the infantry into enemy trenches, then pull the galloping horses around so the guns pointed towards the enemy, unhitch the horses quickly and then open fire at a close distance. Since the enemy had machine guns waiting for such an action, this was not without heavy losses on the part of the field artillery. So tactics had to be changed. This was done as a result of the war itself, which turned from one of motion to one of position. Now artillery on both sides fired from hidden positions only. Artillery had turned into the new master of the battlefield, dictating to the infantry what to do: telling the enemy's infantry to stop, and their own when to advance with the support of the guns. What the artillery conquered in terms of terrain was only later occupied by the infantry. This was overdoing it a bit the other way, but it reflected the war of positions. And it was the reason behind the need for new guns with longer ranges and more effect.

The dugouts had started as simple niches in the forward trench walls facing the enemy. They resembled those in the *Armierungsstellungen*, the prepared field fortifications, in the German fortresses such as Metz, Strasbourg, and so on, built in concrete in peacetime for men, guns and ammunition. These *Unterschlupfe*, refuges, were now a little deeper in the ground, as they were protected only by a cover of earth rather than concrete. The 3m of ground above them was believed to protect against field artillery of up to 10.5cm, so for combating them at

*No longer did the guns (here a 15cm sFH 93) fire close together on the open field ...*

least 15cm calibre at least was needed, in the high-angle mode. Later, the dugouts were made deeper and a second exit provided, a result of the terrible experiences of soldiers buried alive in their shelters. So the *Unterschlupf* became an *Unterstand*, an underground home, no longer forcing the inhabitants to crouch under its 1.5m (4.5ft)-high ceiling, but allowing them to stand as well. Enemy artillery fire hammering on this for days could harm them only by the direct hit of a heavy calibre.

Most soldiers were safe inside, with only a few on guard in the trench, waiting for the enemy to attack. But their warning was not needed; the lull in the continuous bursts of the impacting shells, the *Trommelfeuer*, the drumming fire, told everyone inside the shelter: 'They are coming!' Hurrying out of the shelter, carrying the heavy MG 08, they took their positions in the trench, partly filled in by the shells, and another enemy attack was stopped before it crossed the barbed wire. Days later the artillery would open up for another try, though the effect was minimal.

French artillery was strongest at an offensive in October 1917 named *La Malmaison*. Over thirty-two days they transported 75,000 tons of ammunition to the front. The fire went on for six days continuously. The guns on the French side numbered 812 field guns, 112 light cannon, 420 heavy field

*... even when the crew attempted to take cover by kneeling (10.5cm lFH 98/09).*

howitzers and 160 heavy cannon. This amounted to 66 light and 99 heavy field pieces per km (0.6 miles), firing a total of 2.5 million rounds by the light and 0.9 million rounds by the heavy field artillery. On the German side only 583 guns, about half the French power, answered them. But in the end the French gained only 6km (3.75 miles) on a front 10km (6.25 miles) wide. And in spite of the tremendous number (3.4 million) of shells fired in preparation, French losses were higher than German ones.

This led to a new form of fire, the *Feuerwalze*, fire roll, the creeping barrage. The attacking infantry was to find shelter behind a dense wall of impacting shells, forcing the defenders to keep down in their trenches and shelters. This creeping barrage of explosive shells was started in the no man's land in front of their own trench, moving forward 100m (300ft) every two minutes, time enough for the infantry to follow, until it had passed the first enemy trench, forcing him down all the time. This was employed for the first time by the British on 1 July at the Somme. Later, the infantry was expected to follow this fire at a distance of only 50–30m (150–90ft) behind. The risk of losses caused by fragments or rounds falling short was to be accepted. (Let us compare this to peacetime range safety: all NATO range safety regulations agree that soldiers have to be 1,000m (3,000ft) from where a shell explodes. This safety distance had been 150–200m (450–600ft) before World War One (smaller calibre, less explosive).) The German army also accepted this close following of the infantry.

In the years between 1916 and 1918 this system of the creeping barrage was perfected. The defence system then consisted of up to four lines, one behind another, so fire was opened on all of them, and also on the positions of the batteries, which had been detected by sound and flash ranging beforehand. Long-range guns fired deep into the hinterland, the terrain behind the front lines, stopping reinforcements from advancing to the front. German artillery fired at the first line with *Minenwerfer* out of their own first trench. Fire was kept on each line for three to four minutes and then advanced.

The last German offensive in the west fired a double creeping barrage. The first was with non-resident gas of green- and blue-cross shells, which had dissolved by the time the infantry reached the object. This was followed by a second barrage of explosive shells.

All of these artillery efforts were unsuccessful in the end, the failure of the French Nivelle offensive in April 1917 even resulting in dangerous open mutiny in the French army. One of the reasons for these failures was that the Germans were hardly surprised by the attack, having been shelled for days on end. The French were able to stop the German thrust at Verdun in 1916 after having lost only a few kilometres. The key was to keep the fire shorter than before, as we saw in the case of Verdun.

Nevertheless, the Allied offensives of 1916 and 1917 were accompanied by barrages lasting for days on end. The reason for this seeming contradiction was that the French artillery had less of the modern quick-fire gun models than the German. This was especially true of the heavy mortars needed to fill in the trenches and press in the shelters. In French artillery, the 220mm (8.8in) mortar model 1880 was used. When this had finally finished firing one round, its German counterpart, the 21cm (8in) mortar of 1910 had fired four.

Calculations made by the 2nd German Army said that in three hours the artillery could destroy the following lengths of ditch:

- one battery of 10.5cm light field howitzers with 800 rounds: a 100m (300ft) ditch
- one battery of 15cm sFH 13 with 600 rounds: a 150m (450ft) ditch
- one battery of 15cm sFH 02 with 400 rounds: a 100m (300ft) ditch
- one battery of 15cm sFH 96 with 300 rounds: a 75m (225ft) ditch
- one battery of mortars (three to a battery) with 325 rounds: a 100m (300ft) ditch.

To destroy an enemy battery the mortars needed 300 explosive shells, the 15cm sFH 500, and the 10.5cm lFH 1,000 rounds. For 1m (3ft) of the front, the army calculated that two medium 17cm or one heavy 25cm mine, the shell for the MW, would suffice.

German tactics had benefitted from the experiences in the east, understanding that surprise was of more importance than long artillery preparation. The French and the other Allies, however, believed in 'the artillery as the destroyer of all', wherefore the preparatory fire of the artillery had to be long. This turned the terrain into a moon-like landscape of craters, which together with rain or a high water table, as in Flanders, made it impossible for the artillery to advance to support further infantry attack. The British with their modern quick-fire guns did not need the same length of time as the French and made consequent use of this in some instances. At Cambrai, for example, tanks and infantry advanced with the beginning of the artillery fire. This attack almost penetrated the German lines.

The successful attacks in the east prepared by only short fire made the German command think about making the surprise perfect by doing away with the adjustment. This was the old method of ranging the target by firing at it, adjusting the range of the gun (not all the guns of a battery fired then, only the *Arbeitsgeschütz*, the work gun) by changing the elevation, and the azimuth according to the comments of the observer: 'more left, too far . . .'.

There are very many factors of both interior and exterior ballistics involved in the way a gun fires. The firing tables gave details of which charge to load and the elevation to take for a desired range. But they did not take into consideration other factors, such as the influence of the weather (air pressure, temperature, wind direction and force) or the individual gun, which due to its wear had lost muzzle velocity compared to the values on which the firing table was based. If all this was considered, one could fire without observation, only by using the map.

Tables of this nature had already been calculated for the APK by the German ballistics expert Professor Cranz in 1915 for the super-heavy flat-trajectory fire. Based on these, a Major Pulkowski worked out new tables for firing without observation. For this, each gun had to fire on a firing range from a surveyed position to a target equally surveyed, with the ammunition to be used later. Measuring muzzle velocity in the way the APK wanted would have brought the same results more easily, but the German industry was unable to supply the equipment needed. This Pulkowski procedure, opposed at first by a lot of artillery officers, was introduced in early 1918. Two great German offensives in March and May of 1918 were started without zeroing in the guns on the targets.

Even more was done to camouflage the preparations. Most batteries were emplaced only a few days before the offensive, at night because of enemy

*Now each gun was alone, hiding behind brushwork (15cm Ringkanone) ...*

*... or a parapet (15cm Ringkanone) ...*

*Tactics, Ranging, Transport and Shelters*

*... or in the woods* (above) *(15cm sFH 93) ...*

*... or behind a ridge* (right) *(21cm Mörser).*

planes, and their tracks ploughed over. Wheels and the hooves of horses were bandaged with straw against the noise. This also went for the transport of ammunition. Camouflage was tested by having their own planes look at the hidden emplacements.

Since observers were not needed, the artillery could open fire at 02.00, followed by an infantry attack at 04.00. Within seventeen days the enemy front was pushed back up to 50km (31 miles) deep and 45km (28 miles) wide. But from March until July 1918 the Germans lost one million men, in comparison to the 700,000 of the allies. They would keep up the fight for another three months, until the revolution which began in the German navy finished all efforts on the front in November 1918.

# RANGING AND OBSERVATION

When the bitter lesson of how unhealthy it was to fire from an open position had been learned, the guns started disappearing, firstly keeping to the other side of a hill, then camouflaging their positions, and later

*Tactics, Ranging, Transport and Shelters*

(Above) *10cm* Kanone *17.*

(Right) *The price for forgetting the rules of concealment was great: getting hit by the enemy.*

firing at dark or during times when vision was impaired by fog, either natural or artificially created by smoke generators.

This modesty was not appreciated on the other side of the front, as observation of enemy batteries became almost impossible. For this purpose each battery had received at the turn of the century a special *Beobachtungswagen*, an observation car, on which was mounted a ladder of 5m (15ft) height, terminating in a platform with a seat for the observer behind the binocular periscope, and an armour shield for protection.

Flash ranging had already been used by the heavy German artillery around the turn of the century. This was later forgotten, as in those days of mobile war the one or two days needed to install the necessary telephone lines was considered to be too long. In addition, fighting took place in daylight, when vision did not permit discovering hidden batteries.

The system was reintroduced with the war of positions, and the *Lichtmesstrupps*, flash-ranging troops, came into being. They did not need special equipment for this job – the telescopes of the guns were sufficient to see the flashes of gunfire, even from

*Tactics, Ranging, Transport and Shelters*

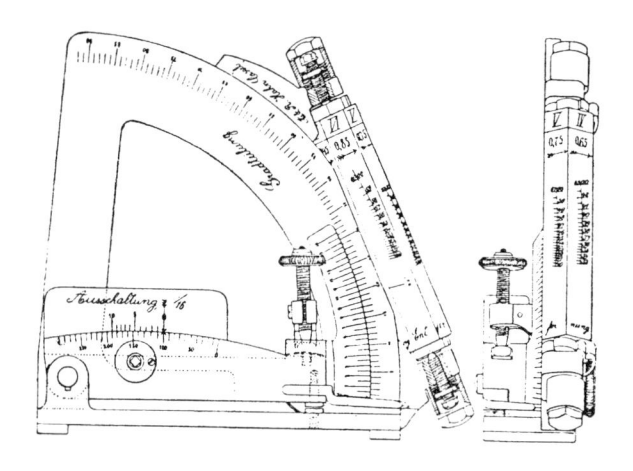

The gunners laid their guns with the help of an *aiming circle* (right) and an *elevation quadrant* (above).

A. Richtkreis
B. Richtkreislager
C. Buchse
D. Visierfernrohr
E. Einheitsgestell

*The Turkish fortifications at Anadoli-Kavak. The medieval forts had been strengthened after 1914 with German help. The shadow of the Zeppelin crossing the area shows from where this excellent picture was taken. It would be hard to improve on, even today.*

hidden guns in daylight, and the gun positions were calculated by triangulation. They also supported their own artillery when zeroing in on a target. In this case the explosion of a few shells fired with time fuses at constant azimuth and elevation, to detonate at a certain height (firing at the enemy target would have given away the planned surprise), were measured and registered on a special map. This took ten minutes and six rounds. Many of the older gunners preferred firing with observation by planes, feeling this *Höhenmessplanverfahren*, height measuring map system, to be too scientific.

Sound ranging took longer to be developed. A German, Dr Leo Löwenstein, requested a patent for his system of sound ranging in November 1913 and submitted his invention to the APK. He was told that his idea would not work, since the speed of sound is not constant, but dependent on air temperature, humidity and wind. This would never work with many guns firing from different positions. Other reasons for the rejection may have included the fact that after the Russian–Japanese war of 1904–5, all artillery experts had been forced to accept that hidden batteries were impossible to discover. The only remedy suggested had been to fire on all parts of the terrain that were likely to hide an observer and thus blind the gunfire. So nothing had been done in this matter when the war came. The same was true for all other armies.

This also changed with the war of positions. In October 1914 *Feldartillerieregiment* 76 ranged a French gun in a forest south of Lille in this way, the first instance of German sound ranging. The ranging centre had installed two microphones right and left, each 2km (1.25 miles) from the centre and behind the front. The observer in the centre held two stopwatches. As soon as he heard the first noise of the shot in his headphones he stopped the first watch; at the second noise he stopped the other one. The site of the gun was then discovered based on the time difference by graphics. This was still an imperfect system, however. One of the possible errors was mistaking the frontal wave of a supersonic shell for the report wave from the muzzle. A mistake of only 0.1 seconds in stopping the watches would place the gun a further 300m (900ft) away. The higher staffs in the army did not take this problem seriously. When a scientist from Zeiss contacted an officer in charge of the *Kriegsvermessungswesen*, war surveying, this Lt Col Boelcke submitted his proposal to the OHL, the supreme army command. They told him to keep his nose out of things that were not his business.

France, with the Bull systems, and especially Britain, with the Tucker microphones which worked with a heated platinum wire, were more open-minded in this matter. It took the German army until 1917 to change its mind, after artillery men complained that precise enemy fire started hitting them after they had fired some forty rounds, showing that both French and British artillery had found their positions quickly. The nine men of the flash and sound ranging detachments, still using the old method with a pair of microphones, were mobilized on carts so that they could follow the advancing attack. Seven of these detachments – four flash and three sound – drove with the attacking German troops in the spring offensive of 1918. Sound ranging now also helped in zeroing in the guns on to a target, just as flash ranging had done.

Observation from the air and aerial photography was used to verify the results of sound and flash ranging. These showed the exact position of the enemy batteries after their approximate position had been found. All three processes complemented one another. Regulations for this were included in Part 5 of the *Vorschriften für den Stellungskrieg* (The War of Positions), *Verwendung und Tätigkeit der Artillerieflieger im Stellungskrieg*, (Artillery Planes), from 1 November 1916.

# TRANSPORT

Artillery was the first large-scale transportation business in warfare and remains so today. We have learned that the guns being transported from one place to another had changed little over five centuries, until the advent of the rifled breech-loader half a century before the war. Transportation had not changed at all.

## Tactics, Ranging, Transport and Shelters

Ever since the days of Sultan Mehmed the Conqueror before the walls of Constantinople in 1453, no one had found a better way than his to save the terrible toil of lugging heavy guns to enemy strongholds: casting them *in situ*, right on the spot where they were to be fired. All other methods started with endless chains of oxen, horses or mules being whipped along, but always finished with sweating gunners straining and cursing to move the guns.

Germany had been an agrarian country, and still was in 1914, despite the onset of industrialization after 1871, the so-called *Gründerjahre*, the years of foundation. Guns were still moved by horses, of which there were enough to go around. They dictated the speed of the troops: 6km/h (3.75mph). And four horses pulled a wagon loaded with 1.2 tons.

This did not change before the war, even when the super-heavy mortars of 42cm calibre appeared. They had been designed to be broken down into loads and then transported on railways – or by steam traction on roads for short distances – to their positions, where they were put together, mounted on their large and time-consuming concrete bedding, and then fired. The German railway had thus helped win the 1870–71 war, and it would help win the next one too. Motorization was slow to develop in Germany, in spite of the motor car having been invented there by Carl Benz, later joined by his assistant Gottlieb Daimler to create Daimler-Benz, today Daimler-Chrysler.

Mechanical traction had started in the German army with steam engines pulling guns and supplies in the 1870–71 war. The idea had originated with Moltke (the older), who bought two Fowler road engines. Steam stayed on afterwards, the heavy road engines with their 41 and 53hp pulling 18 and 24 tons respectively at a speed of 5–7km/h (3–4.5mph). It was also used to pull the loads of 42cm Gamma-mortar or the 15cm *Kanone in Schirmlafette*, the 6in fortress guns of Metz, planned to be moved from one position to another either by railway or by street transport, sitting on steel trailers and pulled by road locomotives. The autocar with combustion engine came later, although already in 1889 Wilhelm Maybach, cooperating with Gottlieb Daimler, had suggested an *Artillerie-Quadricycle*. The idea of combat motor vehicles, proposed by General Richter in 1907, of field guns with armour shields accompanying the infantry, and Burstyn in 1912, of a motorized armoured gun on tracks, remained unheard.

In 1902 the Prussian Ministry of War, together with the Ministry for Agriculture, had advertised a reward for a truck that could transport a load of 10 tons with two trailers cross-country. Daimler and NAG offered ordinary tractors for this purpose, but Siemens came up with a new idea. The four wheels of 2m (6ft) diameter were all driven by electric motors in their hubs, which received the current from a gasoline-driven generator on the car. This car also consisted of two halves, moving independently of each other – the first articulated drive. In 1906 a road train of a similar but simpler type was ordered from Siemens by the army, transporting 15 tons on five trailers, all driven by an electric engine on the rear axle. But the road trains caused too much wear on the roads and were of poor off-road capability, so they were rejected in Germany. (But not in Austria, where the Porsche *C-Zug*, the C-train, developed from the *Landwehr-Train*, was to move even the heaviest guns in the war.)

So the search for vehicles suited to the needs of the army went on. There was a special department in the Prussian Ministry of War: the *Abteilung 7 Verkehrstruppen*, which undertook the necessary trials, testing ordinary trucks for their usefulness for the purposes of the army. This A7V also developed, after a long period of experimentation, special army trucks with trailers that together transported 6 tons at a speed of 8–12km/h (5–7.5mph) – a lot for the roads of that time. Money was paid to firms that bought the types of trucks that the army considered suitable, which would have to be handed over to the army in times of war. In 1911 there were 611 of these truck trailers.

In 1913 the APK, who already saw all guns pulled either by horses or by mechanical traction, asked the KM, the War Ministry, for motor traction for the transport of 30.5cm Beta-mortars as well as the coming 42cm *M-Gerät* and the 15cm KiSL guns. In 1914 the APK went for a light tractor for pulling 9 tons and a heavy one for 24 tons. For the latter, in

1914 Podeus came up with an improved motor plough that had 80hp. Of these, seventeen were ordered for further trials, and Podeus had to agree not to sell this type to foreign countries. Four-wheel drive was considered to be a waste of money, since the artillery was supposed to stay on the roads. The only exemption from this doctrine was granted to the *Ballonabwehrkanonen* (BaK), designed for chasing after the balloons and dirigibles. The later makeshift solution of about 150 Belgian 5.7cm casemate guns mounted on ordinary four-ton trucks for antitank fire was just this – makeshift.

When war broke out in 1914, the VPK, the transport commission, bought sixty steam plough engines and eighty-seven steam road engines from civil firms, and before May 1916 another forty-three. This was the end of steam traction in Germany.

From July 1914 onwards motor tractors running on benzene were tested for pulling the different loads of *Beta-* and *M-Gerät*. At first the number of breakdowns disappointed the VPK and APK and steam traction was considered again for a time, but by the end of the year this was definitely declined. In March 1915 an Ing. Bräuer had proposed to the APK a load-distributing system for heavy guns of his invention, which lowered the ground pressure by suspending the gun between the tractor and a sort of semitrailer. The APK accepted this and Rheinmetall went on to produce it. At the end of 1915 there was agreement that motor traction was needed for all guns of the *Fußartillerie*, not only the super-heavy guns. This led to more Bräuer load-distributors being ordered, for the 13cm guns and the new 15cm guns model 16. The normal army trucks of 35hp were not strong enough for these loads, so new artillery tractors of between 60 and 80hp were demanded by the APK and developed by the industry. On 13 July 1916 the APK ordered forty tractors with 100hp for the 42cm *M-Gerät*, and sixty with 80hp for the 15cm *Kanone* 16. In November 1916 the AD (5) demanded a special truck for ammunition transport and ordered 200 of them for 1917.

Krupp had an agreement with Daimler about a tractor for the 15cm K 16 Krupp, that resulted in a special *Krupp-Daimler-Artillerie-Zugmaschine*, the KD I. The design was solely agreed by these two monopolists, without any input from the military, neither the APK nor the VPK – a strange procedure. The KD I was finished in 1916, tested on training ranges and supplied to the troops in 1917. The order was for 1,070 KD I for pulling 15cm K 16 and 740 for pulling antiaircraft guns, together with some for the 17cm L/40 ex-naval gun; of these, 981 were delivered by the war's end.

On 24 November 1916, KM laid down the official designations of all military motor vehicles:

- *Artillerie-Kraftschlepper*, tractor, 80hp, with load-distributor for 15cm K 16 (Rheinmetall), 21cm mortars and similar.

*The first standardized military truck was the* Artillerie-Kraftzugmaschine *KD I, built by Krupp and Daimler. It was to pull the new 15cm* Kanonen *16 of Krupp and Rheinmetall.*

## Tactics, Ranging, Transport and Shelters

*The Krupp-Daimler* Artilleriezugmaschine *KD I. The steel wheels can be equipped with* Radgreifern, *wheel grippers.*

(Left) *The craters made by shells in roads and terrain made transport of ammunition a difficult business. The best off-road ability was shown by vehicles laying their own tracks and so the chassis of the German tank A7V was also used as a* Überland-Geländewagen, *a cross-country vehicle. The armour was replaced by two cargo platforms front and rear, and the vehicle used to transport artillery ammunition or to tow guns.*

- *Artillerie-Kraftzugmaschine*, tractor, with 100hp, for pulling super-heavy and 15cm K iSL
- *Artillerie-Kraftzugmaschine* Krupp-Daimler, tractor KD, with 100hp, for medium guns, especially the Krupp version of the 15cm K 16
- *Dampfzugmaschine*, steam engine, with 80hp, for 15cm guns and heavy coast mortars.

\* The number of horsepowers was always added to the designation, for example Art.Kr.Zugm. 100 PS.

Further and stronger *Zugmaschinen*, the Z II and Z III, were developed later in 1917 by Daimler, again without asking the military, and only according to Krupp's demands. Krupp insisted that they had a contract guaranteeing them the sole authority on such vehicles for the heavy artillery.

There was less friction in the field of transport for light and medium artillery. The three firms of Magirus, Dürrkopp and Büssing had submitted designs of four-wheel drives – Daimler, Lanz and Benz vehicles – with tracks. The winning Magirus-type had 70hp and a winch for pulling itself out when stuck. It was introduced as the *Kraftprotze* 1 (KP 1), the power limber 1, in the middle of 1918. A planned successor, KP 2, with articulated frame

210

*The 15cm* Kanone *16 of Krupp was transported in two loads: tube* (top) *and carriage* (above).

*The 15cm K 16 by Rheinmetall was transported in one load with the help of a* Lastenverteilergerät, *load distributor, an invention by Bräuer.*

*The* Lastenverteilergerät *could also move the* lange *21 cm* Mörser *16.*

steering was not finished before the end of the war. Other motor limbers tried and ordered by the German army were eighteen for 1.6 tons with an air-cooled 16hp engine from the Austrian Daimler company at the end of 1917, and two from the Austrian Fiat company.

On 15 July 1918 the park for artillery tractors at Opladen held 387 *Zugmaschinen*. Of the Z III, seventy-six were ordered. They were destined for pulling the five loads of each of the twelve new 24cm (9.6in) guns ordered for April 1918. Altogether, at the end of October 1918, the artillery had ordered 3,049 tractors, received 2,078, and was waiting for another 1,034.

On 15 November Daimler and all other firms working for the Wumba, the German arms and ammunition procurement office, received a telegram from the office which read: *nr 4920 – reduction as fast as possible without dismissing personnel, speedy conversion to peacetime production desired. Further details to follow – Wumba 425 11 18 3 6.* This was the end.

Germany had lived up to her usual motto for all wartime inventions: too late and not enough. Thus German motor transport never reached the level of that of the Allies on the western front. During the last years of the war the Allies had about 100,000 trucks, used for transporting troops and ammunition. In comparison Germany had about one third of this number, only 35,000, of which only 23,000 were trucks. These were facing the problems caused by lack of rubber for tyres since 1917. A typical solution was cast-steel wheels running on steel rims, with inlays of wood for 'spring'. Other

*If the tractor got stuck, it could anchor itself and use the winch to pull a trailer.*

## Tactics, Ranging, Transport and Shelters

*A better solution was offered by the* Räder-Raupen-Fahrzeug, *wheel-track-vehicle, of which the* Kraftprotze, *the motor-limber of Benz-Bräuer was the first example. This vehicle was either a half-track with tracks on, or a wheeled truck without them.*

*Another solution was to lengthen the pedrails to a sort of track, shown here on a 21cm experimental mortar L/12 by Rheinmetall.*

designs at least used coil springs between wheel and rim for this purpose. Of course, these wheels could only run at limited speeds (12km/h, 7.5mph), still sliding around on the roads, ruining both themselves and the vehicles. The frame and cardan shaft suffered from the vibrations so much that the drive of the rear axles had to be reconverted to the old chain drive system. It was of little help that the four-cylinder gasoline engine of the KD I delivered 100hp, which could speed the KD I up to 35km/h (22mph) on the road. Offroad it put extra *Radgürtel*, pedrails, on the rims of the wheels to lower ground pressure.

Attempts at German selfpropelled guns for infantry support had also been made, with the APK demanding a mobile armoured gun platform for FK 16 and lFH 98/09. They were to be mounted on tracked vehicles of Lanz and the *Marienwagen*, and on motor-limbers of Benz, Bräuer and Büssing construction. But none were introduced except for the earlier motorized Bak-/Kw-antiaircraft guns.

The horse was also the workhorse, growing weaker as the war went on and his fodder became less and poorer (on the western front in 1918 German horses received only 6.3lb oats or barley instead of their daily ration of 12.6lb, and only one third of the planned hay and straw). At the same time the guns had become heavier – in the case of the FK 16 rising to 4,830lb on the march, compared to the 4,011lb of the older FK 96 n/A. But *Kamerad Pferd* remained a faithful comrade until his death. Of the 1.27 million horses in the German army during the war, no less than 68 per cent were lost.

## HARDENING THE GUN POSITIONS

Until 1900 the trend in German artillery had been toward a light field gun without armour shield, old generals showing their contempt for the 'cowards' hiding behind the shield mounted on their gun. The famous French 75mm M 97 not only changed opinion in Germany concerning long recoil, but also as to whether it was smart to risk losing a well-trained guncrew to infantry fire. So the FK 96 n/A was also a better field gun than its older sister FK 96 (later also called *alter Art*, old style), because it too received an armour shield. With other German field guns the protection varied. The 10.5cm light field howitzer had a shield; the heavy 15cm of 1902 did not, but got one in 1913; the 10cm *Kanone* of 1904 also did not have a shield, but got one in the 1914 model. Both 15cm guns of 1916 had one, whilst the 21cm mortar of 1910 did not, but received one in the 1916 model. It seems that the experiences of the war changed a lot of minds.

But these shields were only in front of the gun, not full all-around armour. Besides the immobilized fortress guns in armour cupolas or *Schirmlafetten* of navy style, the latter in 15cm calibre also transportable, there was no such protection. The only known guns to be clad in armour later on were some of the 42cm *Gamma-Mörser*, the older heavier model in this calibre, firing from a concrete bedding (not to be confused with the *M-Gerät* of the same calibre, the one and only *Dicke Bertha*).

On the other hand, German artillery did believe in protecting the guns. All prepared defence positions around the fortress towns, such as Strasbourg, Metz, Königsberg, and so on, had not only the *I-*, *A-* and *M-Räume* concrete shelters for soldiers of infantry, of artillery and for the ammunition, but also concrete gun positions, where the guns, mostly old reserve cannon 9cm C/73/91, stood sheltered inside during enemy fire and were wheeled out into the firing position behind a parapet. The *Fahrpanzer* in the fortifications were fighting with the same principle: stay under cover until the enemy artillery has to stop firing because their troops are attacking, then come out and fire as fast as you can. In 1909 the less glorified *Fußartillerie*, foot artillery, of the heavy siege guns received a regulation for the protection of their batteries, the *Batteriedeckungs-Vorschrift für die Fußartillerie*.

The war of positions in the west now forced the field guns to take as much protective cover as possible. Like the infantry with their *Unterschlupf*, dugout, the underground shelter was the obvious

*Tactics, Ranging, Transport and Shelters*

*Here a 42cm Skoda mortar is assembled. The bedding platform has been laid, the trailer with the carriage driven onto the bedding, and now the tube will be inserted into the cradle.*

*(Below) The 42cm Austrian mortar. This one has received some extra armour to protect mortar and crew, just as some of the German 42cm Gamma-mortars did.*

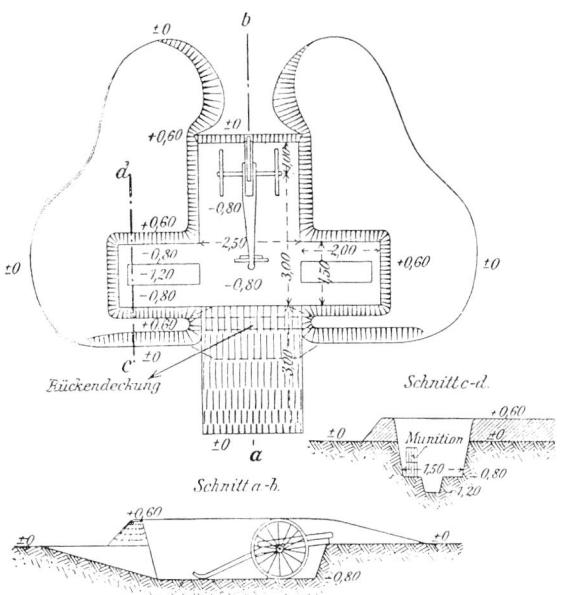

*Before the war the regulations had already foreseen two different types of cover for the gun positions of the German field artillery. The preparation of one had been ordered when there was time enough to do so: the prepared position. This even provided cover against shells detonating in the rear of the position.*

*Tactics, Ranging, Transport and Shelters*

(Right) *Less time and effort could be devoted to cover during a battle that suddenly developed. But even in this case the crew of the gun was to do something for their own survival, and especially that of the gun. This was to consist of: (a) digging one spade deep into the ground, (M) trenches for the men and the crossed area showing where sandbags, in this case bags filled with the soil gained by digging out (a) and (M), were to be placed to the left of the gun and to the right between the gun and the limber with the ammunition.*

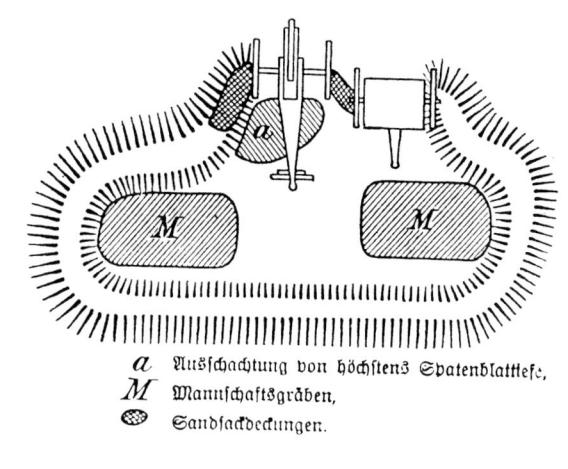

*a* Ausschachtung von höchstens Spatenblatttiefe.
*M* Mannschaftsgräben,
⊗ Sandsackdeckungen.

answer. But this only enabled high-angle weapons to fire from it. It was fine for the *Minenwerfer*, but not for the flat trajectory field artillery. Only in the stopping positions built some kilometres behind the front lines were concrete gun positions already prepared, providing safety against all but the heaviest enemy fire, but not against gas. Some of them already looked a lot like the later gun casemates of *West-* and *Atlantikwall*. The rules for building these and other positions were laid down from 1916 onwards in a series of regulations known as *Sammelheft der Vorschriften für den Stellungskrieg für alle Waffen* (Collection of Regulations for the War of Positions for All Branches of the Service). Almost full protection against enemy fire for the German guns came with the armour of the tanks, but for the small calibre of 57mm only.

(Above) *Protection from the enemy's artillery was life insurance, as seen here for the infantry who built the dugouts in their trenches in wood. Note the periscope, which consists of one half of a Scherenfernrohr.*

*Even more protection was offered further in the rear, where the artillery was positioned, by using reinforced concrete. The old Sütterlin writing on the back of the wartime field postcard speaks of 'an indestructible thickness of 5m'.*

216

# 7   The End in 1918

The war of 1914–18 was characterized by the artillery. The war of 1870–71 was fought with one gun for every 350 soldiers; when World War One began this ratio had increased to one gun for every 200 soldiers, and kept growing, reaching one gun for every 60 soldiers between 1916–18, not counting the *Minenwerfer*. The thirty-six guns of a German *Feldartillerieregiment* were able to fire as many shells within three months as the whole German artillery in the 1870–71 war: 670,000 rounds. The artillery had turned into the decisive branch of World War One, in terms of both cost and effect.

Germany entered the war in August 1914 with 9,388 field and 1,396 heavy guns (in total almost the same number as the *Wehrmacht* on 1 September 1939); during the battles on the west front in July 1918, Germany had 13,100 field and 7,300 heavy guns. Against these the Allies massed, in the same period, 19,804 field and 8,323 heavy guns.

At the end of the war, the number of guns of the German artillery amounted to 19,808 field and 7,860 heavy guns. And now for the surprise: according to the dictate of Versailles, Germany was forced to hand over or destroy exactly 59,897 guns, more than twice as many as the German army had had at the end of the war. Another case of *Vae victis*, woe to the defeated? Of course it was, but it could still be done. So where did all these extra guns come from?

The total number of guns produced by Germany during the war was much higher than indicated by the number left at the end of the war. One reason for this was enemy action. One single bullet of an infantryman's rifle or a plane's machine gun was sufficient to put a gun out of action, if this bullet did not hit the massive tube but the thin walls of recoil brake or recuperator. The gun then had to go back, in this case not to the factory but to a weapon maintenance unit in an old factory behind the front, where the damaged parts were exchanged. Guns were also worn out by firing. The number of rounds it took for this to happen depended on calibre, charge and also rate of fire, with a tube heated by long or rapid firing wearing out earlier. The extreme case in this regard were the antiaircraft guns of 8.8cm and 10.5cm. Field artillery lived a bit more healthily and therefore longer, starting with the 20,000 rounds survived by the smaller calibres, down to the 2,000 of the 15cm guns of 1916. Field guns firing at an average daily rate of thirty-five rounds had a life expectancy of eighteen months, until they had to be sent back and replaced, this time to their place of birth at Krupp. There the tube was restored to life, either by replacing the whole tube as in the case of the FK 96 n/A, or by pulling out the worn inner liner and pressing a new one into the outer tube. This meant that a lot of guns were recycled.

The heavy German artillery saw an increase in guns from 1,396 to 7,860 during the war. It also saw an increase in personnel. In 1914 it consisted of 1,420 officers, 33,250 NCOs and men, and 3,400 horses. In 1918 there were 18,500 officers, 400,000 NCOs and men, and 202,500 horses. The last number rose disproportionately because of the need for transport for the highly increased ammunition rates, and also because the poorer quality of horse fodder meant that the animals pulled less weight, so more of them were required for the same load.

## The End in 1918

During the war the German artillery fired millions of rounds:

- 7.7cm FK: 156 million
- 10.5cm FH: 67 million
- 15cm FH/K: 42 million
- 21cm M: 7 million
- all calibres: 272 million rounds.

Collectively, the armies of the nations involved in the war fired 856 million rounds at an estimated cost of thirty thousand million dollars. Between 1 January and 11 November 1918, the British army fired 71.5 million rounds and the French army 81 million. The rate of fire rose tremendously at the end of the war, with France firing about 50 per cent of their ammunition used during the whole war in the last ten months, and Britain 58 per cent.

The amount of explosive fired by shells rose even more. In the German army this was 650 tons per month in 1914, rising to 10,000 tons per month for the shells of artillery and *Minenwerfer* and for grenades. The numbers wounded by this fire rose at the same rate too. Before the war, the French calculated losses by rifle fire to be four times those by artillery, but found when actually evaluating the battles that 67 per cent of the losses had been caused by fragments of shells fired by artillery, *Minenwerfer* and grenades, 23 per cent by rifle and machine gun fire, and 10 per cent by other causes. During the war of positions in the west in 1917, the Germans registered 75 per cent of the wounds to have been caused by artillery. This shows that the war was dominated by the artillery (together with the machine gun). Its fire killed a vast number of humans on both sides and destroyed whole cities. Yet it still did not bring a victory for any side; it just slowly ground them both down.

This chapter could end on this solemn note, as we have now covered all German artillery of World War One. But have we? Everyone with an interest in history could contradict this statement by pointing out that not all guns have been covered: those from the end of the war seem to be missing – those of 1918 vintage, characterized by their name ending in '18'.

Examples of these include the 10.5cm *leichte Feldhaubitze* 18 and the 15cm *schwere Feldhaubitze* 18, the 8.8cm Flak 18, the 21cm *Mörser* 18, and so on, all of them also fielded later by the artillery of the *Wehrmacht* in World War Two. Yet all of these bear the number '18' unjustifiably. They lie about their age deliberately. They were designs of the 1920s right after the war, a period when the development of arms by Germany was forbidden. However, it still continued in neutral countries like Switzerland and Sweden. The number '18' was meant to camouflage the identity of these weapons.

The armistice on 11 November 1918 decreed that:

- Germany had to retreat from all occupied areas at once, and out of Alsace-Lorraine within 15 days;
- the peace treaties with Russia and Roumania of 1917 were no longer valid;
- no German soldier was to remain within 30km east of the Rhine;
- Germany had to hand over 5,000 artillery guns, 25,000 machine guns, 5,000 railway engines and 150,000 railway cars at once;
- Germany had to hand over its complete navy and all submarines;
- the hunger blockade against Germany would continue.

Note that all of this had to be performed before the peace treaty at Versailles was even started.

The dictate of Versailles of 29 July 1919 had not only forbidden Germany most types of weapons, such as submachine guns, machine guns, planes, tanks or heavy or antiaircraft guns, but had also forbidden their development. German fortifications were destroyed in the west along the French border some 50km (31 miles) deep, and forbidden in the east along the new border with Poland (which the Polish tried to shift even further to the west in 1923, until the old soldiers of the *Freikorps* rallied, digging up their *Gewehr* 98 and MG 08/15, and driving them back until 1945). The new army, the *Reichswehr*, was limited to only 100,000 men, a mere drop in the ocean compared to the number of its enemies. Almost all arms had to be handed over to the victors, who then kept them or even sold them for profit, thus often furthering the war between neighbouring countries.

Besides part of her terrain in the east, Germany was forced to give up:

- 59,897 guns and 31,470 *Minenwerfer*, together with the tubes;
- more than 6 million rifles;
- almost 40 million artillery rounds and 491 million cartridges;
- 16,000 planes, fighters and bombers, with 28,000 aircraft engines;
- 26 battleships, 23 cruisers, 315 submarines, 83 torpedo boats, and 21 training ships;
- 2,500 machine tools of the armament industry; the rest were destroyed.

Was this a sound basis for everlasting peace, or did it cry out for revenge, as France had cried in the years after 1871, until it was finally heard?

All of these actions were monitored closely by a special committee of the allies: the *Interalliierte Militärkontrollkomission* (IMKK), Interallied Military Control Commission. Meanwhile, the rest of the world continued to develop and produce modern arms. This was certainly not satisfying, even for the new democratic German government, ruling after the Kaiser had abdicated and been exiled to Holland. So the *Reichswehr* and the industrial sector found ways of circumventing the embargo. The big manufacturers such as Krupp and Rheinmetall cooperated with foreign weapon makers or started a clandestine development hidden from the eyes of the IMKK. This led in the 1920s to new weapons such as the MG 13, 15 and 17, the aforementioned field howitzers and antiaircraft guns with an '18' in their name.

All new guns, no longer dating back to World War One, were ready in time for World War Two, only twenty-one years later. Mankind seemed unable to learn, in spite of all the millions killed or mutilated, the suffering, the mourning of families who had lost husbands and fathers, the destruction of the ground that had been battlefields, the senseless waste of money for this.

Looking back gratefully at the fifty-five years of peace we in the West have enjoyed since the end of World War Two, I would like to thank those to whom we owe this happy time: to my comrades, to all the soldiers of NATO, who have watched silently and continually all this time, when the world was often on the brink of the next war, this time definitely the last one. We were prepared then, and let us stay this way: *Si vis pacem, para bellum*, If you want peace, be prepared for war.

# Glossary of German Terms

**a/A, alter Art** Old model. When a gun or shell, for example, was later improved, the designation for the old model.
**A7V; Abteilung 7 Verkehr** Transport Department of the Prussian War Ministry; also the name of the German tank A7V developed by it.
**Artillerieprüfungskommission (APK)** Artillery Testing Committee.
**Armee** (1) Army in general; (2) a unit consisting of several corps.
**Artillerie** Artillery.
**Aufschlagzünder** Impact fuse either with or without delay.
**Ballonabwehrkanone** Antiballoon gun (old name for antiaircraft gun).
**Bak** Short for *Ballonabwehrkanone*.
**Bertha** First name of Krupp's daughter, later given by the soldiers to the 42cm mortar *M-Gerät*.
**Bismarck, von** Chancellor of Germany, who kept a delicate peace both with Britain and Russia. Dismissed by Wilhelm II, whereupon Russia joined France, surrounding Germany.
**Beta-Mörser** 30.5cm (12in) mortar.
**Brennzünder** Powder train fuse.
**C/?; Construction of ?** Model of gun, shell, and so on, named after the year of introduction. In use until about 1900.
**Dicke Bertha** Fat Bertha (*see* **Bertha**).
**Doppelzünder** Dual fuse, MTSQ fuse.
**DWM** *Deutsche Waffen- und Munitionsfabrik* in Berlin.
**Eberhardt, von** Chief ballistician at Krupp.
**Ehrhardt** Founder, owner and chief designer of Rheinmetall.
**Eisenbahngeschütz** Railway gun.
**Ersatz-** Substitute.
**Fahrpanzer** Mobile armour housing with 53mm Gruson gun.
**Falkenhayn, von** General, replaced von Moltke as Chief of General Staff. Lost at Verdun and was replaced by von Hindenburg.
**Feldhaubitze** Field howitzer, firing in low and high trajectory.
**Feldkanone** Field cannon, firing in low trajectory only.
**Flak** *Fliegerabwehrkanone*, antiaircraft gun.
**Fliegerabwehrkanone** (*see* **Flak**).
**Flügelmine** Fin-stabilized bomb fired from smoothbore mortar.
**Gamma-Mörser** 42cm mortar, firing from concrete bedding.
**Gebirgsgeschütz** Mountain gun, always broken down into several loads.
**Gewehrprüfungskommission (GPK)** Rifle Testing Committee.
**Geschoss** All sorts of projectiles: armour-piercing shells, illuminating shells, and so on.
**Granate** Explosive shell.
**Granatenwerfer** Light mortar of spigot-type, also called *Priesterwerfer*.
**Granatzünder** Fuse for explosive shell.
**Gruson** Owner of an iron works in Magdeburg; famous for his armour made from hard cast iron; helped Schumann realize his ideas; invented 5cm quick-fire gun. Taken over by Krupp in 1892.
**Hindenburg, von** Defeated the Russians at Tannenberg, replaced von Falkenhayn and was later German *Reichspräsident* until 1934.
**Kaliberlänge** L/?, length of a tube, expressed in multiples of the calibre.

## Glossary of German Terms

**Kanone** Gun, cannon, firing in flat trajectory mostly, except for antiaircraft guns.
**Kriegsminister** Secretary of War.
**Krupp** Iron works in Essen on the Ruhr, founded by Alfred Krupp, who developed his famous cast steel and went into gunmaking, where he rose to be number one in the world.
**L/?** (*see* **Kaliberlänge**).
**Ladung** Charge; guns fire either fixed cartridges or different charges.
**Leicht** Light, as in *leichter Minenwerfer*.
**Lettow-Vorbeck, von** General, led the *Schutztruppe* in the German East Africa.
**Marine** Navy.
**Matrosenartillerie** The heavy ex-naval guns fielded on land were served by naval gunners.
**M-Gerät** 42cm mortar on a wheel carriage; M- is for *Minenwerfer*.
**Messei** Crusher gauge, used for measuring gas pressure in tube.
**Minenwerfer** Mine launcher, short-range mortar of the engineers.
**n/A, neuer Art** New model. When a gun or shell, for example, was later improved, the designation for the new model.
**Oberste Heeresleitung (OHL)** Supreme command of the army.
**Panzerkuppel-** Armour cupola.
**Panzerturm-** Armour turret, rotating 360 degrees.
**Parisgeschütz** Paris gun, also Wilhelm gun: a 21cm long-range gun, also rebored to 23.2cm. Also designated as *lange* 21cm *Kanone in Schiessgerüst*.
**Rausenberger, Professor** Chief constructor/designer at Krupp.
**Rheinmetall** Formerly *Rheinische Metallwaarenfabrik* at Düsseldorf; rival of Krupp; founded by Ehrhardt.

**Schirmlafette, SL** The protective armour box of the 10cm and 15cm fortress guns. The guns were then 10cm/15cm KiSL guns in SL.
**Schlieffen, von** Chief of German General Staff 1891–1905; foresaw a war on two fronts, and planned for this by attacking France first through less fortified Belgium: the *Schlieffenplan*.
**Schnelladekanone** Around 1900, name of modern quick loading guns firing charges contained in metal cartridge cases with primer.
**Schumann, von** Engineer Officer, who after retiring cooperated with Gruson in realizing his ideas on armour fortification.
**SK** (1) *see* **Schnelladekanone**; (2) later *Schiffskanone*, ship gun.
**Spandau, Berlin** Centre of Prussian armament, with rifle, gun and ammunition factories and construction offices.
**Sturmabteilung** Special units formed at the end of the war for attacking.
**Tankabwehrkanone (Tak)** Antitank gun.
**Verkehr-, V-, VPK** Traffic-, Transport-; name of department in KM.
**Wilhelm II, Kaiser** German Emperor 1888–1918; grandson of Queen Victoria; was declared a war criminal in 1918 and his extradition demanded by the Allies and declined by Holland.
**Wilhelmgeschütz** (*see* **Parisgeschütz**).
**Wumba** *Waffen- und Munitionsbeschaffungsamt*, Office for Procuring Arms and Ammunition.
**Zeitzünder** Time fuse, either mechanical (clockwork) or powder train.
**Zeppelin, von** Invented the rigid airship, the dirigible named after him.
**Zünder** Fuse.
**Zugmaschine** Prime mover, tractor.

# Index

2cm Becker aircraft gun 91, 94, 162
2cm Becker antiaircraft gun 91, 94
2cm Becker antitank gun 144
2cm Becker M II 148
2cm Ehrhard aircraft gun 161, 164
2cm Ehrhard antitank gun 144, 148
3.7cm Bak of 1870 87
3.7cm Flak 95
3.7cm GrabenK 134
3.7cm KasemattK 94
3.7cm LandungsG Gruson 90, 180
3.7cm LandungsG Maxim 90, 179
3.7cm *Luftschiff Flak* Kr 92, 180
3.7cm Maxim MK 90, 92
3.7cm RevolverK 90, 91, 101
3.7cm RevolverK i.Wall L. 97
3.7cm *Sockelflak* 92
3.7cm SturmbegleitK 135
3.7cm TankG 160
3.7cm Tak Fischer 144, 148
3.7cm Tak Krupp 147
3.7cm Tak Rheinmetall 147
3.7cm Tak, *verbesserte* Rh 152–3
5cm KasK 99, 101
5cm K i.Pz.L. 98, 99
5cm Bak L/30 Rh 87
5cm Flak Rh 95
5cm Tak 153

5.2cm KüstK 114
5.2cm TankK 160
5.7cm IG L/30 139
5.7cm IG 143
5.7cm KasK (B) 143, 145
5.7cm TankG 157
6cm K i.SenkTurm 99
6.5cm Bak L/35 87
7.5cm FK 97 (F) 19, 89, 93
7.5cm FK 06 (I) 94
7.5cm FK 11 (I) 94
7.5cm GebK 05 Rh 86
7.5cm GebK 08 83
7.5cm GebK 13 84
7.5cm GebK 14 84
7.5cm GebK 15 (R) 84
7.5cm GebK M15 (Sk) 84, 138
7.5cm IG L/20 139
7.5cm Kw Flak 87
7.5cm lMW *DoppelLauf* 150
7.62cm FK 00 (R) 93
7.62cm FK 00 Rh 93
7.62cm FK 02 (R) 93
7.62cm GebK 09 84
7.62cm IG L/16.5 136
7.62 KasK (R) 136
7.7cm Bak 88
7.7cm C-Geschoss 117
7.7cm FGr 96 191
7.7cm FK 96 15
7.7cm FK 96 a/A 15, 115
7.7cm FK 96 n/A 17, 19, 32, 115
7.7cm FK 16 117
7.7cm FSchr 96 191
7.7cm GebK M15 Rh 84
7.7cm IG L/19.5 138
7.7cm IG L/20 137

7.7cm IG L/27 137
7.7cm IG 18 139
7.7cm KanGr 14 195
7.7cm KasemattG 102, 159
7.7cm KGr 15 mP 152
7.7cm Kw Gesch 14 145, 150
7.7cm NahkpfG 142
7.7cm NahkpfG M17 139
7.7cm PanzerK 159
7.7cm L/27 *Sockelflak* 88
7.85 lMW a/A 71
7.85 lMW n/A 72, 134
7.92mm × 57 cartridge 144, 149
8cm KwFlak 92
8mm MG 08 101, 143
8mm Mondragon rifle 161
8mm twin-MG Gast 150, 161
8.8cm Flak 92, 95
8.8cm Kw Flak 96
8.8cm S.K. L/35 114
9cm K C/61 8, 9, 10, 12
9cm K C/64 10
9cm K C/73 13, 14, 90, 178
9cm K C/73/88 13, 14
9cm K C/73/88/91 13, 14
9cm K C/79 20
10cm GebH M16 (Sk) 85
10cm K 02 Rh 21
10cm K 02/03 Kr 21
10cm K 04 17, 22, 85
10cm K 04/12 22
10cm K 14 23, 85, 121
10cm K 17 121, 122
10cm K 17/04 122
10cm K 77 (R) 33
10cm K C/99 21

10cm KiSL, *Festung* 100
10cm KiSL, *Marine* 106, 179
10cm KüstK L/50 107
10cm SK L/40 114
10cm TurmK 100
10.5cm FH Gesch 05 199
10.5cm FK L/35 (*zerlegbar*) Rh 85
10.5cm Flak Kw/E 93
10.5cm GebH 12 Kr 84
10.5cm GebH 12 Rh 84
10.5cm K L/35 Rh 23
10.5cm Kw Flak 93
10.5cm lFH 98 17, 20
10.5cm lFH 98/09 85
10.5cm lFH 16 94, 117
12cm K M62 (B) 9
12cm sK C/80 24
12 in How L/18.8 (GB) 133
13cm K 09 60
13mm cartridge 143
13mm MG TuF 145, 150, 162
13mm antitank rifle Mauser 145
15cm KiSL 61, 62, 117
15cm le Kartaune Rh 118
15cm lg RingK C/92 25
15cm lg RingK 24
15cm Pressluft MW 75
15cm RingK C/72 18
15cm RingK C/92 25
15cm sFH C/93 17, 25
15cm sFH 02 17, 26, 86
15cm sFH 13 26
15cm sFH 13 L/17 27
15cm sFH 13/02 28
15cm sFH 07 Rh 28

# Index

15cm sFH 14 Rh  29
15cm sK 16 Kr  118, 119
15cm sK 16 Rh  118, 119
15cm SK L/30 iRL  63, 121
15cm SK L/45 E  133
15cm TurmH  100
15cm VersH Rh  120
17cm m MW a/A  71
17cm m MW n/A  71
17cm SK iSL  112
17cm SK L/40 E  64, 126
17cm SK L/40 iRL  64, 65, 132
18mm Dreyse needle rifle  8, 12
18.5cm *Vers.Haubitze*  123, 124
21cm lg Kan i. *Schiessgerüst see* Paris Gun
21cm *der Mörser* L/12  37
21cm *Mörser* C/99  34
21cm *Mörser* 07 Rh  38
21cm *Mörser* 09 Rh  38
21cm *Mörser* 10  18
21cm *Mörser* 16  121, 124
21cm Paris Gun *see* Paris Gun
21cm SK L/35  105
21cm SK L/40  66
21cm SK L/45  114
21cm SK L/45 E  133
21cm TurmH  98
21cm *Versuchsmörser* 07  36, 38
21cm *Versuchsmörser* 09  38
21cm *Wilhelm-Geschütz* L/162 *see* Paris Gun
23cm lg K i. SchiessGer  176
23.2cm WilhelmG L/146 *see* Paris gun
24cm sFlügelMiW  75
24cm SK L/30 E  132
24cm SK L/40 E  127
25cm s MW a/A  69
25cm s MW n/A  70
26cm Pressluft MW  75
28cm Küstmörser  114
28cm *Mörser* L/12  42, 105, 106
28cm *Mörser* L/14  42, 44

28cm RingK L/22  104, 107
28cm RingK L/30  104
28cm SK L/40 E  112, 128
28cm SK L/45  132
30.5cm H E (Gb)  133
30.5cm H iRL L/17  40
30.5cm H iRL L/30  58
30.5cm Mrs Beta  39
30.5cm Mrs Beta 09  39, 40
30.5cm Mrs Beta-M L/30  58, 59
30.5cm SK L/50  105, 113
35.5cm SK L/53  67, 125, 129, 167
35.5/38 SK L/45  67
38cm SK L/45 E  67, 109, 125, 130
38cm s schw MW  75
40.6cm SK/KüstK  59, 166
42cm M-Gerät  36, 39, 50–7
42cm Mrs Gamma  39, 46, 47, 48
42cm Mrs iRL  36, 39, 50–7
42cm Mrs (Sk)  58, 215
60/54cm *Mörser* KARL  58
80cm *Kanone* (E) DORA  45, 132

A7V, detachment  155, 208
A7V, tank  155
A7V, transporter  210
acoustic camouflage  174
aerial observation  207
aircraft guns  161
Altenwalde  171
ammonium nitrate  195
antiaircraft guns  87
AA ammunition  198
antitank guns  141
APK  9, 14, 16, 18, 49, 87, 117, 143, 149, 152, 208
armistice  217
armour cupola  44, 98
armour piercing  146, 152, 197
Armstrong-guns  12, 62, 64
*Aufschlagzünder*  22, 192, 196

automatic guns  161

*Ballonabwehrkanone see* Bak
Bak  88, 95
balloon bombs  45
barrage fire  198
base bleed, BB  176
Becker  59, 91, 114
Behelfs-Minenwerfer  75, 76, 77, 81
Belgium  34
*Beobachtungswagen*  205
Berlin  8, 118
Berlin-Spandau *see* Spandau
*Beta-Mörser*  39
*Beta* 09-*Mörser*  39, 40
*Bettungsschiessgerüst*  165
Bismarck, von  11, 19, 22, blackpowder  9, 62, 187
breechloader  7–9, 39, 188
Brialmont  34, 97
Bosch  75
Bräuer  209, 211
Büssing  155
Burstyn  155

C/42  7
C/61  8, 9, 10, 12, 104
C/64  10
C/73  13, 14
C/73/88  13, 14
C/73/88/91  13
camouflage  58, 112, 174, 203
canister  191
*Carbonit-Werke*  163
casemate guns  7
Cavalli  7, 8
Clausewitz  23, 35, 105
coast artillery  18, 104
creeping barrage  202
crinoline mount  145, 157
crusher gauge  175
C-shell  117

Daimler, Gottlieb  208
Daimler, plant  118, 212
Dardanelles  177
dead ground  189

*Dicke Bertha*  39, 55, 57, 165
*Doppelzünder*  193
Dreyse needle gun  8
DWM  144
dugouts  200

East Africa  86, 178, 180
Eberhardt, von  176
Ehrhardt  16, 68, 148, 161
*Einheitsgeschoss*  190
*Eisenbahngeschütze*  125
engineer-committee *see* Ingenieurkomitee
*Erdmörser*  75
explosives  218

*Fahrpanzer*  15, 98, 101
Falkenhayn, von  182
*Feldgranate*  96
*Feldhaubitze*, FH  17
*Feuerwalze*  202
fin stabilized  101
Fischer  144
FK 96  15
FK 96 a/A  15
FK 96 n/A  17, 32
*Flachfeuer*  59, 165
flash ranging  205
*Flügelminenwerfer*  75–8
Fokker, Anthony  161
Fort Camp des Romains  180
Fort Douaumont  185
Fort Genicourt  180
Fort Loncin  39, 57
Fort Pontisse  36, 51
Fort Vaux  184
fortress artillery  96
Fowler steam engine  154
*Füllpulver* 02  49, 72
*Füllpulver* 60/40  195
fuses  192

Galopin turret  184
*Gamma-Mörser*  48, 50, 56
gas pressure  174
gas shells  183, 191, 197
*Geschützfabrik*, Spandau  11
*Gewehrgranate*  16  80

223

# Index

*Grabenkanone* 134
*Granatenwerfer* 1 75
*Granatfüllung* 88 14, 49, 191
*Grosskampfwagen see K-Wagen*
Gruson 14
Gruson guns 83
Gruson hard-cast iron 15, 98
Gruson plant 15
Gruson revolver gun 97
guns, models of 1918 33, 218

*Halbrundkeil* 12, 13
*Handgranate* 79
*Haubitzgranate* 14
*Haubitzenzünder* 14
Haussner, Konrad 14, 16
high angle 39
Hindenburg, von 182
horses 214, 217
Hotchkiss-Gruson 97
howitzers 17

impact fuse 192
IMMK 219
*Infantriegeschütze* 133
infantry guns 133
*Ingenieurkomitee* 49, 68, 101

*Kampfwagen, leichter* 157, 159
*Kanonengranate* 14 195
*Kanonenzünder* 14 195
*Kartaune, leichte* 118
  *schwere* 58
KD I 209
*Keilverschluss* 45, 188
*Kraftprotze* 210
*Kraftwagen* (Kw) 92
Krupp, Bertha 39
Krupp, Dr Gustav, von Bohlen-Halbach 36
Krupp cast steel 9, 11
Krupp-Daimler 209
Krupp factory 18, 49, 50, 87
*Kugelhandgranate* 13 80

*K-Wagen* 156, 159

L/... 18, 188
*Lafettenschwanz* 52
*Landungsgeschütz* 179
Lanz 210
*Lastenverteilergerät* 40, 61, 211
Layritz, von 154
lead coat, thick 187
  thin 187
Liège/Luik/Lüttich 35
long recoil 16, 68
low-angle fire 59, 165
Ludendorff, von 35, 167
Luger, Georg 144

M 61 11
M 62 9
M 97 16, 19
Maas-mill 182
Malandrin plate 115
MAN 150
*Marienwagen* 214
*Marineinfantrie* 179
Marne miracle 31, 181
*Matrosenartillerie* 108, 173
Maybach, Wilhelm 208
*M-Gerät* 50
*Minenwerfer, leicht* 71, 134
  *mittel* 71
MW 49, 68
  *schwer* 50, 68
motor traction 208
Mougin 97, 98, 125
mountain guns 82
*Munitionswagen* 30
muzzle loader 7, 39, 188
muzzle velocity 175

Napoleon III 11
Nordenfeld gun 150

observation 204
*Obus Robin* 190
OHL 75, 91, 141, 164, 181

*Panzerlafette* 99
*Panzerkopfgranate* 146, 197

Parabellum-MG 144
Paris 31, 165
Paris Gun 164–76
pedrail 48, 124, 213
picric acid 14, 18, 49, 191
piston breech 8
pressure gauge 175
*Priesterwerfer* 75, 78
primers 192
prismatic powder 64
propellant 66, 168, 171, 174, 192

quick-fire guns see *Schnell-adekanone*

*Radgürtel see* pedrail
rail guns 133
railway artillery 125, 132
ranging 204
Rausenberger, Professor 50, 164
recoil system 14, 173
Rheinmetall 16, 17, 49, 149
rifle grenade 16 80
road train 154
Rogge, Vizeadmiral 66

*Schirmlafette* 130
Schlieffen, von 22, 31
Schmeisser 144
*Schnelladekanone*, SLK 66
*Schrapnell* 191
Schumann 15, 97
Schumann turret 97
shells 189
shelters 200, 214
shrapnel 191
SK 66
SLK see *Schnelladekanone*
smokeless powder 174
sound ranging 207
Spandau 11, 90
spin stabilized 8
split trail 28
steam traction 51
*Stielhandgranate* 79, 142
Stokes, Sir Wilfrid 69, 79
*Sturmabteilung* 136
*Sturmbegleitkanone* 135

*Sturmpanzerwagen* 155
*Sturmwagen, kleiner* 160
superheavy mortars 34

tactics of artillery 31, 200
tanks 141
Tak see antitank gun
*Tankabwehrkanone see* Tak
tank guns 154
time fuse 193
TNT 72
tracked vehicles 210, 213
transport 207
trench gun 134
trench mortar 75
trinitrotuluol see TNT
*Trommelfeuer* 201
Tsingtao 105
tube length see L/...
tube life 217
Tucker microphone 207
TuF 145
turtledove 78

Uchatius bombs 45, 163
Uchatius bronze 20

van Essen, Pieter 190
Vécer 75
Verdun 180, 186
  battle of 180
  citadel 180
  forts 180
Versailles, dictate 218
Vollmer, Hptm 155
VPK 155

Wahrendorff breech system 7, 8, 9
Wahrendorf, von 7, 8
wedge breech 10
Wilhelm II 6, 22, 219
*Wilhelm-Geschütz* 164
Wumba office 212
*Wurfmaschine* 75
*Wurfmine* 74, 78, 79

*Zeitzünder* 193
zero in 7
*Zeughaus*, Berlin 8
*Zündnadelgewehr* 8

224